LIMNOLOGY - SOME NEW ASPECTS OF INLAND WATER ECOLOGY

Edited by **Didem Gökçe**

Limnology - Some New Aspects of Inland Water Ecology
http://dx.doi.org/10.5772/intechopen.73416
Edited by Didem Gökçe

Contributors

Marcos Nogueira, Juliana Pomari, Kenneth Cummins, Felipe De Lucia Lobo, Evlyn Márcia Leão De Moraes Novo, Claudio Clemente Faria Barbosa, Vitor Hugo Fernandes De Vasconcelos, Cleber Mansano, Luiz Sérgio Vanzela, Juliana Heloisa Pinê Américo-Pinheiro, Beatrice I. Macente, Kifayat U. Khan, João Batista K. Fernandes, Danila Fernanda Rodrigues Frias, Marta Verardino De Stéfani, Noureddine Bouaïcha, Naila Yasmine Benayache, Tri Nguyen-Quang, Fatima-Zohra Mehennaoui-Afri, Kateryna Hushchyna, Benjamin Onozeyi Dimowo, Adewole Gbadebo, Adewale Taiwo, Isaac Omoniyi, Didem Gokce

Notice

First published in London, United Kingdom, 2019 by IntechOpen
IntechOpen is the global imprint of INTECHOPEN LIMITED, registered in England and Wales, registration number: 11086078, The Shard, 25th floor, 32 London Bridge Street
London, SE19SG – United Kingdom
Printed in Croatia

British Library Cataloguing-in-Publication Data
A catalogue record for this book is available from the British Library

Additional hard and PDF copies can be obtained from orders@intechopen.com

Limnology - Some New Aspects of Inland Water Ecology, Edited by Didem Gökçe
p. cm.
Print ISBN 978-1-83880-788-7
Online ISBN 978-1-83880-789-4
eBook (PDF) ISBN 978-1-83880-790-0

For EU product safety concerns:
IN TECH d.o.o., Prolaz Marije Krucifikse Kozulić 3, 51000 Rijeka, Croatia,
info@intechopen.com or visit our website at intechopen.com.

Meet the editor

Dr. Didem Gökçe is an associate professor of Biology at Inonu University, Malatya, Turkey. She graduated with a BSc in Biology from Hacettepe University, Ankara, Turkey where she also obtained her MSc and PhD degrees in Hydrobiology and Limnology. Her research focuses on water quality, limnology, and ecosystem ecology. Dr. Gökçe works on combining experimental and field studies to qualitative and quantitative population and community structure on plankton and benthic macroinvertebrate ecology.

Contents

Preface

Limnology is a multidisciplinary science that covers biological, physical, and chemical properties of lakes, rivers, and other waters of the inland aquatic ecosystem. Limnology has many important aspects. These aspects are categorized into three main groups: (1) assessing the quality and quantity of biological communities in the freshwater ecosystem, (2) determining the physical and chemical properties, and (3) as a result of these evaluations, establishing and applying a water management plan on the basis of limnological data.

The problems of worldwide freshwater resources (e.g., water pollution, climatic change, salinization, decreasing water level, and loss of freshwater biodiversity) have become a severe limitation for the concept of the future ecosystem continuum. Freshwaters are fragile and vulnerable ecosystems. That is why limnology is a subject with an ever-increasing importance. The purpose of this book is to discuss contemporary limnological problems by putting a special emphasis on freshwater biota and application of management techniques to monitor water quality.

Limnology: Some New Aspects of Inland Water Ecology is comprised of seven chapters written by renowned experts from Algeria, Brazil, Canada, France, Nigeria, Turkey, and the United States. These authors provide updated information on taxonomy, ecology, methodology, and remote sensing. I would like to thank all of them for their invaluable contributions to this book.

Chapter 1 presents the importance of limnology and some of today's problems (nanoparticles, microplastics, climatic change, etc.) affecting inland water ecosystems. Chapter 2 discusses the occurrence of cyanobacterial Harmful Algal Blooms (cyanoHAB) as well as environmental factors favoring their proliferation, possible human and animal health outcomes associated with their toxins, and a review of robust modeling approaches to predict the bloom pattern. Chapter 3 mentions that Diatom-based indices are increasingly becoming important parameters for the assessment of ecological conditions in lotic and lentic ecosystems. Chapter 4 explains the use of the functional feeding groups (FFG) method, which is not similar to any traditional use of taxonomic analysis, in order to gain a rapid and efficient insight into macroinvertebrate community composition and its function in freshwater ecosystems. Chapter 5 provides information about the main activities to develop a monitoring system for quantifying water siltation caused by small-scale gold mining in the Amazon rivers using multi-satellite data. Chapter 6 presents the main limnological patterns of a tropical/subtropical reservoir aiming to provide scientific protocols for management purposes and a long-term water quality monitoring program in the Paranapanema River reservoir cascade (southeastern Brazil). Chapter 7 focuses on the importance of water quality for "raniculture" and its development protocols, as physical and chemical parameters of water should be considered and analyzed together because all of these factors have a direct impact on the culture systems.

I would like to offer special thanks to Author Service Manager Jasna Bozic for her help in bringing out this book in its present form. I am also grateful to IntechOpen for their concern, efforts, and encouragement in the task of publishing this book.

Dr. Didem Gökçe
Department of Biology
Faculty of Science and Letters
Inonu University
Malatya, Turkey

Section 1

Introduction

Chapter 1

Introductory Chapter: Current Status of Freshwater Ecosystems

Didem Gökçe

Additional information is available at the end of the chapter

http://dx.doi.org/10.5772/intechopen.86628

1. Introduction

Limnology covers biological, chemical, and physical qualifications of lakes and other bodies of inland water as a whole ecosystem. It has become a predominant sub-branch of ecology due to aquatic limiting environmental factors, population dynamics, and community structure. Limnology is the term derived from the Greek word limne—lake and pond. Although limnology initiated with lake studies, it has expanded its field with the studies on rivers, wetlands, and estuarine ecosystems.

Wetzel [1] clearly expresses that freshwater ecosystems are biological systems, and biology controls water quality under natural conditions, and freshwaters must be evaluated and managed as biogeochemical systems. Each freshwater ecosystem maintains its balance under the current conditions (community composition and physicochemical factors) depending on its carrying capacity. Anthropogenic factors can change these conditions negatively. Therefore, engineering solutions are not sufficient alone. It is crucial to know the quality and functions of aquatic ecosystems in order to manage them successfully [1–3].

In that case, physical and chemical changes in habitat conditions lead to impairment in water quality. The physicochemical variable analysis is a classic method of controlling pollution and managing water quality. The community shows a habitat-specific distribution depending on its carrying capacity, and thus biological monitoring reflects the balance of that aquatic ecosystem within the cause-effect relationship [4]. Therefore, a water management tool has been developed by using bioindicators to evaluate the effects of the ecosystem quality on aquatic organisms [4–6]. Preserving water quality is a critical part of sustainable management of water basins. It affects the health of the hydrological system. A healthy system will provide better water quality and a more flexible ecosystem.

2. Problems of inland waters

Today, along with the advancement of technology, the property of pollutants given to the environment has started to become more destructive. The effects of combined environmental pollutants cause the ecosystem to deteriorate more rapidly. Many environmental factors can affect the community structure, distributions of organisms, and energy transfer between trophic levels in inland aquatic ecosystem. Anthropogenic activities such as agriculture, domestic and industrial wastewater discharges, and natural factors (geomorphology, hydrology, seasonal variations, and climate) change the physicochemical quality of water ecosystems.

2.1. Nanoparticles

Being used in a rapidly increasing rate in the world, the engineered nanoparticles (NPs) are widely present in many products when entering the aquatic ecosystem. Most of NPs are converted into free metal ions in solution mainly depending on the particle size, surface area, and rough degree (such as $npTiO_2$ and npZnO) [7]. For this reason, NP residues are present in numerous waste materials, and these nanoparticles have a potential risk for freshwater ecosystems [8, 9].

Nanoparticles TiO_2 ($npTiO_2$), zinc oxide (npZnO), cerium oxide ($npCeO_2$), copper oxide (npCuO), and silver (npAg) are the most studied metal and metal oxide nanoparticles which have toxic effects on freshwater biota and food web dynamics. Even though the NP toxicity studies on freshwater organisms (algae, zooplankton, and fish) were carried out on different experimental models, the impact of food web on NP toxicity is still not clear [8–12]. The behaviors of NPs in the aquatic system include aggregation, flocculation, redox reactions, speciation, dissolution (release of + ions like Zn^{2+}), surface modifications (interactions with other metal and pollutants), and complexation with natural organic matter.

The size of nanomaterials has a direct and critical effect on the mechanical and physiological activities in aquatic organisms [8]. Especially NPs are absorbed during filtration, hence have an impact on feeding ability in *Daphnia magna* and plays a crucial role in cellular uptake. Nanoparticles can change the community composition at the level of species that are more sensitive or tolerant to environmental contamination. Gokce et al. [8] explain that as a result of the effect of increasing NP concentrations, population growth rate reduces depending on the delay of progeny, delay of the reproduction period, and mortality rate. Taken together, these results have revealed that NPs exposure may display a negative effect on community dynamics of aquatic organisms and on food chain structures in freshwater ecosystems particularly for a long period.

2.2. Microplastics

Nowadays, plastics have a wide usage area due to increasing population. The environmental impacts first began to be noticed in the seas in the 1970s. While in terms of sea and seashore pollution the effects of plastics have been studied intensively, their effects on freshwater ecosystems and organisms have attracted the attention of researchers in recent years. Different criteria can be considered in the grouping of plastics: size (micro-, 1–5 mm; meso-, <5 mm; and macroplastics, coarse structure), primary and secondary plastics, and density, low- and high-density plastics. Plastics are suspended in water or accumulated in sediment according to these properties [13].

The damage/hazard to the aquatic ecosystem varies according to the properties of plastics. Plastics indicate accumulation in organisms by feeding on water column or sediment. They have a mechanical and chemical effect on living things. They also cause organic and metal pollution.

Due to the formation of large specific areas of microplastics and their hydrophobic structure based on their small size, they cause the accumulation of organic matters and metals.

Fish is the most studied group in freshwater ecosystems. It has been defined that microplastics have accumulated in the gill, digestive system, and even muscle tissues. As a general effect, it has been found that they cause differences in the immune system and changes in leukocyte levels and oxidative stress levels [13–15]. It has also been observed in invertebrates that it accumulates in the digestive system by feeding group. Different levels of microplastic accumulation are found in benthic invertebrates according to the way of feeding. *Daphnia* species, which are widely used as experimental model organisms, have been detected in the digestive system as well as the clutch. In acute and chronic assays, it has been found that immobilization increases and survival time and reproduction decrease [16].

Different types and concentrations of microplastics have been identified in different lakes and streams. The microplastics have different sources: the primary and secondary plastics (such as private care products and the wastewater from the treatment facilities located in the vicinity). As a result, anthropogenic pollution accumulates in the water body and sediment. In addition, as a result of absorption of different chemical groups along with changing of environment conditions (temperature, pH, salinity, etc.), formation of the different pollution load depending on the microplastic nature causes the deterioration of water quality in the freshwater ecosystem. It is inevitable that this pollution load will be transported to fish, waterfowl, and human beings via the food web.

2.3. Medical wastes

The widespread emergence of resistance to pharmaceutics among pathogens has become one of the most critical issues worldwide. Hospital wastewater contains a group of organisms that cause various diseases, and it plays a major role in the spreading of drug-resistant pathogens by becoming a pollutant by improving the growth, propagation, and warming properties in the environment. Especially wastewaters that are involved in lakes and rivers cause very important problems [17]. These conditions have become a major problem for public health, particularly in developing countries as they can be transmitted directly or indirectly to human beings and animals from the aquatic food web [18].

Due to the excessive increase in the human population, the quality and quantity of pollutants in inland water ecosystems have changed. One of these contaminants is pharmaceuticals.

Pharmaceuticals are emerging as aquatic ecosystem contaminants, and their concentrations in water bodies are an increasing environmental issue [17–20]. Although their positive effect on health (human, veterinary, water culture, etc.) is significant, antibiotic-resistant bacteria and genes are found in freshwater environments, and their transfer to the food chain causes numerous negative effects. Different antibiotic pharmaceuticals are crucial environmental micro- and nano-pollutants, and their presence in freshwater is a serious environmental problem. Several antibiotics have been reported in freshwater habitats. This is parallel to the global increase in antibiotic consumption [19].

The detection of pharmaceutics in treated wastewater was first reported by Fent et al. [21] in 1976 in Kansas City, USA. Twenty-five different pharmaceutical concentrations above 1000 ng L^{-1} in the Lee River (England) were recorded in 1981. Globally, there is antibiotic contamination in freshwater habitats despite the presence of an advanced waste treatment plant in the USA and European countries. The situation is much more serious in the middle- and low-developed countries [19].

Several studies have been carried out on freshwater and sediment-receiving wastewater contaminant with pharmaceutics. In a recent study by Varol and Sünbül [20], different levels of 37 antibiotics, 10 metals, and 19 organochlorine pesticides were analyzed in 6 fish species from the Karakaya Dam Reservoir (HEPP), Turkey. Fish is one of the groups used as bioindicators for the evaluation of water quality and pollutants such as antibiotics and heavy metals and are taken directly from the water column and trophic transfer [17, 18, 20]. Therefore, for human health, it is very important to determine the levels of these pollutants in consumable fish species because the consumption of these fish species is the main way of exposure to pollutants.

Appropriate management of hospital wastewater should be implemented to reduce the problem at each healthcare facility. Therefore, the mixing of pharmaceuticals and hospital wastes with the increase in surface water, which is the receiving medium, should be prevented.

The ozonation, chlorination, and UV disinfection used independently or in combination in the disinfection of medical and pharmaceutical wastes are among the most investigated subjects and have demonstrated several performances. The use of coagulation technology to reduce antibiotic-resistant genes in wastewater treatment plants has been found to be more effective [17]. Moreover, graphene-based TiO_2 composite photocatalysts are efficient methods for the removal of antibiotics, antibiotic-resistant bacteria, and genes from wastewaters [17]. In addition, as a result of administration of TiO_2 photocatalysis under UV irradiation in combination with H_2O_2, good removal efficiencies of antibiotic-resistant bacteria and antibiotic-resistant genes (both intracellular and extracellular forms) from aquatic media were reported.

2.4. Climatic change and the related effects

Water temperature is an important environmental factor that limits survival, growth, and reproduction of plants and animals. Plant and animal populations are more likely to survive if they enter fields with climatic conditions similar to their natural habitats. The increase in water temperature and salt concentration and even pH change due to global warming cause the invasive species to spread to new habitats. The higher evaporation rates with the decrease in precipitation cause the water level to decrease, and thus most of the water ecosystems are under threat.

Climate change is a substantial factor affecting species invasion, biodiversity, and aquatic ecosystem structure in freshwater. The impact of climate change on freshwater biodiversity is increasing the speed of biological invasions [22]. Changes in climatic conditions, population dynamics of the indigenous species, and the composition and structure of communities can change the functioning of ecosystems. Similar to the observed responses of indigenous species, climate change can directly affect the possibility of invasive species entering a different aquatic ecosystem [22, 23].

Especially, salmonid fish and cold stenothermic macroinvertebrates are expected to disappear from many central and southern European river systems. In central and northern European countries, a general increase has been observed in a number of Mediterranean Odonata species, and African Odonata species are also expanding into southern Europe. However, the dispersal of Euro-Iberian species narrowed [24, 25].

The increased water temperature has caused significant changes in the European benthic invertebrate community. It is estimated that approximately 5–20% of invasive species have potent effects on receptor environments [24]. Endemic taxa will be under threat as a result of both losses of habitat and reduced connection between habitats, especially when water flow connections are broken.

Therefore, it is of great importance to reduce the anthropogenic effects of global climate change, which leads to warming and acidification of freshwater in the world. Prevention of nonindigenous species from entering into new habitats is of critical significance.

Algal bloom and eutrophication are one of the huge results of climate change. Furthermore, algal bloom cause increasing water temperature and nutrients, and deterioration of natural and/or human-induced ecosystem balance. Especially in lakes and streams, the decrease in water level and the abundance of toxic algae species are among the important problems faced by today's inland water ecosystems. Particularly, Cyanobacteria and the other harmful algae bloom cause contamination of drinking water sources and aquaculture and can harm the economy and human health.

As a result, many lakes, rivers, wetlands, and estuaries can be affected by climate change, eutrophication, water level decreases, biodiversity reduction, and invasive species. For this reason, water management plans should be prepared. An important deficiency in most of the regions investigated in limnology studies is that freshwater management plan and strong connections between the institutions should be implemented by the decision-making institutions on limnological concepts and sustainable plans. Efficient and practical adaptive measures should be taken at the local scale, both limnologically and economically.

3. Conclusions

The quality and quantity of pollutants have also changed depending on the advancing technology and industry. Considering the rapidly decreasing freshwater resources, the magnitude of the danger also arises. The accumulation of pollutants in the water ecosystem and the aquatic food web results in magnified concentrations of pollutants.

The existence of an increasing number of conflicting freshwater uses is a consequence of the increased anthropogenic pressure on the freshwater ecosystem balance and carrying capacity. At this point, the importance of limnology emerges once. Index methods developed on investigation of the community structure and dynamics of the aquatic organisms (spatial and temporal variations) and long-term monitoring allow the effects of pollutants on the ecosystem to be reliable.

It is a disastrous picture. However, it is observed that global freshwater resources are getting worse in the future. Knowledge about water, its management, and use will continue to grow and evolve in the future. The fact that limnology is a multidisciplinary science enables the determination of ecosystem questions from a wide perspective and the preparation of realistic management plans. Therefore, it is important for decision-makers to form and plan water quality standards on the basis of the principles of limnology.

Author details

Didem Gökçe

Address all correspondence to: didem.gokce@inonu.edu.tr

Department of Biology, Inonu University, Malatya, Turkey

References

[1] Wetzel RG. Freshwater ecology: Changes, requirements, and future demands. Limnology. 2000;**1**:3-9

[2] Mahendra R, Indarchand G, Avinash P, Biswas JK, Sinitsyna OV. Nanomaterials: What are they, why they cause ecotoxicity, and how this can be dealt with? In: Rai M, Biswas JK, editors. Nanomaterials: Ecotoxicity, Safety, and Public Perception. Switzerland: Springer Nature Publisher; 2018. pp. 3-18. DOI: 10.1007/978-3-030-05144-0_1

[3] Altenburger R, Brack W, Burgess RM, Busch W, Escher BI, Focks A, et al. Future water quality monitoring: Improving the balance between exposure and toxicity assessments of real-world pollutant mixtures. Environmental Science Europa. 2019, 2019;**31**:12. DOI: 10.1186/s12302-019-0193-1

[4] Gokce D. Wetland importance and management. In: Gokce D, editor. Wetlands Management—Assessing Risk and Sustainable Solutions. London: Intech Open Publisher; 2019. pp. 1-8. DOI: 10.5772/intechopen.82456

[5] Boggero A, Fontaneto D, Morabito G, Volta P. Limnology in the 21st century: The importance of freshwater ecosystems as model systems in ecology and evolution. Journal of Limnology. 2014;**73**:1-3. DOI: 10.4081/jlimnol.2014.948

[6] Sultana J, Recknagel F, Tibby J, Maxwell S. Comparison of water quality thresholds for macroinvertebrates in two Mediterranean catchments quantified by the inferential techniques TITAN and HEA. Ecological Indicators. 2019;**101**:867-877. DOI: 10.1016/j.ecolind.2019.02.003

[7] Bundschuh M, Seitz F, Rosenfeldt RR, Schulz AR. Effects of nanoparticles in fresh waters: Risks, mechanisms and interactions. Freshwater Biology. 2016;**61**:2185-2196. DOI: 10.1111/fwb.12701

[8] Gokce D, Köytepe S, Özcan İ. Effects of nanoparticles on *Daphnia magna* population dynamics. Chemistry and Ecology. 2018;**34**:301-323. DOI: 10.1080/02757540.2018.1429418

[9] Iswarya V, Palanivel A, Chandrasekaran N, Mukherje A. Toxic effect of different types of titanium dioxide nanoparticles on *Ceriodaphnia dubia* in a freshwater system. Environmental Science and Pollution Research. 2019;**26**:11998-12013. DOI: 10.1007/s11356-019-04652-x

[10] Falfushynska HI, Gnatyshyna LL, Ivanina AV, Khoma VV, Stoliar OB, Sokolova IM. Bioenergetic responses of freshwater mussels *Unio tumidus* to the combined effects of nano-ZnO and temperature regime. The Science of the Total Environment. 2019;**650**:1440-1450. DOI: 10.1016/j.scitotenv.2018.09.136

[11] Bhuvaneshwari M, Iswarya V, Chandrasekaran N, Mukherjee A. A review on ecotoxicity of zinc oxide nanoparticles on freshwater algae. In: Rai M, Biswas JK, editors. Nanomaterials: Ecotoxicity, Safety, and Public Perception. Switzerland: Springer Nature Publisher; 2018. pp. 191-206. DOI: 10.1007/978-3-030-05144-0_10

[12] Gagné F, Auclair J, Turcotte P, Gagnon C, Peyrot C, Wilkinson K. The influence of surface waters on the bioavailability and toxicity of zinc oxide nanoparticles in freshwater mussels. Comparative Biochemistry and Physiology, Part C. 2019;**219**:1-11. DOI: 10.1016/j.cbpc.2019.01.005

[13] Li J, Liu H, Chen JP. Microplastics in freshwater systems: A review on occurrence, environmental effects, and methods for microplastics detection. Water Research. 2018;**137**:362-374. DOI: 10.1016/j.watres.2017.12.056

[14] Rezania S, Park J, Din MFM, Taib SM, Talaiekhozani A, Yadav KK, et al. Microplastics pollution in different aquatic environments and biota: A review of recent studies. Marine Pollution Bulletin. 2018;**133**:191-208. DOI: 10.1016/j.marpolbul.2018.05.022

[15] Lu K, Qiao R, An H, Zhang Y. Influence of microplastics on the accumulation and chronic toxic effects of cadmium in zebrafish (*Danio rerio*). Chemosphere. 2018;**202**:514-520. DOI: 10.1016/j.chemosphere.2018.03.145

[16] Liu Z, Yu P, Cai M, Wu D, Zhang M, Huang Y, et al. Polystyrene nanoplastic exposure induces immobilization, reproduction, and stress defense in the freshwater cladoceran *Daphnia pulex*. Chemosphere. 2019;**215**:74-81. DOI: 10.1016/j.chemosphere.2018.09.176

[17] Lorenzo P, Adriana A, Jessica S, Carles B, Marinella F, Marta L, et al. Antibiotic resistance in urban and hospital wastewaters and their impact on a receiving freshwater ecosystem. Chemosphere. 2018;**206**:70-82. DOI: 10.1016/j.chemosphere.2018.04.163

[18] Mao D, Yu S, Rysz M, Luo Y, Yang F, Li F, et al. Prevalence and proliferation of antibiotic resistance genes in two municipal wastewater treatment plants. Water Research. 2015;**85**:458-466. DOI: 10.1016/j.watres.2015.09.010

[19] Fekadu S, Alemayehu E, Dewil R, Bruggen B. Pharmaceuticals in freshwater aquatic environments: A comparison of the African and European challenge. The Science of the Total Environment. 2019;**654**:324-337. DOI: 10.1016/j.scitotenv.2018.11.072

[20] Varol M, Sünbül MR. Environmental contaminants in fish species from a large dam reservoir and their potential risks to human health. Ecotoxicology and Environmental Safety. 2019;**169**:507-515. DOI: 10.1016/j.ecoenv.2018.11.060

[21] Fent K, Weston AA, Caminada D. Ecotoxicology of human pharmaceuticals. Aquatic Toxicology. 2006;**76**:122-159. DOI: 10.1016/j.aquatox.2005.09.009

[22] Hesselschwerdt J, Wantzen KM. Global warming may lower thermal barriers against invasive species in freshwater ecosystems—A study from Lake Constance. The Science of the Total Environment. 2018;**645**:44-50. DOI: 10.1016/j.scitotenv.2018.07.078

[23] Kiesel J, Gericke A, Rathjens H, Wetzig A, Kakouei K, Jähnig SC, et al. Climate change impacts on ecologically relevant hydrological indicators in three catchments in three European ecoregions. Ecological Engineering. 2019;**127**:404-416. DOI: 10.1016/j.ecoleng.2018.12.019

[24] Gourault M, Petton S, Thomas Y, Pecquerie L, Marques GM, Cassou C, et al. Modeling reproductive traits of an invasive bivalve species under contrasting climate scenarios from 1960 to 2100. Journal of Sea Research. 2019;**143**:128-139. DOI: 10.1016/j.seares.2018.05.005

[25] Robarts RD, Zohary T. Limnology and the future of African inland waters. Inland Waters. 2018;**8**:399-412. DOI: 10.1080/20442041.2018.1481729

Section 2

Some New Aspects of Freshwater Ecology

Chapter 2

An Overview of Cyanobacteria Harmful Algal Bloom (CyanoHAB) Issues in Freshwater Ecosystems

Naila-Yasmine Benayache, Tri Nguyen-Quang, Kateryna Hushchyna, Kayla McLellan, Fatima-Zohra Afri-Mehennaoui and Noureddine Bouaïcha

Additional information is available at the end of the chapter

http://dx.doi.org/10.5772/intechopen.84155

Abstract

This chapter will present an overview of cyanobacterial harmful algal blooms (cyano-HABs) and biotic and abiotic factors, as well as various aspects associated with these worldwide ecological bursts. The exact causes of the cyanoHABs are still not well defined, but eutrophication and climate change (temperature increase, light intensity variation, etc.) are the two assumed main factors that may promote the proliferation and expansion of cyanobacterial blooms. However, these premises need to be profoundly investigated as the optimal combination of all factors such as increased nutrient loading, physiological characteristics of cyanobacterial species, and climate effects which could lead to the blooming pattern will require robust modeling approaches to predict the phenomena. Negative issues associated with cyanoHABs are diverse including the toxic products (cyanotoxins) released by certain taxa which can damage the health of humans and animal habitats around the related watershed as well as generate a huge water quality problem for aquatic industries.

Keywords: cyanobacteria, cyanotoxins, freshwater ecosystems, mathematical modeling, ecotoxicity

1. Introduction

Freshwater ecosystems (lakes, rivers, and reservoirs) play an important role in regulating Earth's climate and they are of high ecological and socioeconomic value, as well as a crucial life-giving resource for humanity. However, these water bodies are fragile and anthropic pressures such as discharge of sewage, industrial pollutants, eroding soil, and deposition of effluents rich in nitrates and phosphorus cumulating by the tourist industry and urbanization have accelerated the rate and extent of their continuous eutrophication which has led to a loss of water quality and biodiversity. Worldwide, the massive proliferation of cyanobacterial harmful algal blooms (cyanoHABs) is among the major undesirable effects resulting of eutrophication [1–4]. To date, environmental factors identified as contributing toward their global expansion included increased nutrient inputs via anthropogenic activities and temperatures and CO_2 concentrations due to changing global climate [5–10]. Nevertheless, these aspects need to be investigated in the future and the combination of increased nutrient loadings, physiological characteristics of cyanobacterial species, and climate change such as increase of temperature and variation of light intensity and quality will require robust modeling approaches to predict the blooming phenomena.

Since freshwater bodies around the world can be used as drinking water reservoirs or recreational areas, the blooming phenomena have gained attention as possible health hazards. The problems associated with toxic cyanobacterial blooms in these different areas are diverse, from environmental asphyxiation due to excessive consumption of oxygen, to purely esthetic problems in recreational areas when the blooms are a colorful and often smelly scum on the surface of the water [5, 11, 12]. To these common problems are added productions by certain species of cyanobacteria of various secondary metabolites with a specific toxic potential (hepatotoxins, neurotoxins, and dermotoxins) causing water quality problems for fisheries, aquaculture, farming, and sanitary hazards to human and animal health [13–17]. This chapter focuses on the cyanoHAB occurrence as well as on environmental factors favoring their proliferation, possible human and animal health outcomes associated with their toxins and a review of robust modeling approaches to predict the bloom pattern.

2. Involving factors in cyanoHAB pattern

2.1. Abiotic factors

Among the abiotic factors, nutrients, including inorganic nitrogen (N) and phosphorus (P), temperature, and light intensity, and hydrodynamic parameters of the water body (turbidity and residue time) have been reported as the most important factors in the proliferation of cyanobacteria [2, 4, 6]. The availability of nutrients such as N and P is essential for the growth of cyanobacteria. For example, field experiments by monitoring surveys of phytoplankton for 3 years (2012–2014) in Lake Erie (U.S.A.) showed that inter-annual differences in the duration, intensity, and toxicity of cyanobacterial blooms in this area were considered related to in-lake and tributary nutrient (N and P) concentrations [2]. This ecosystem observation is consistent with other earlier field and laboratory studies which have shown that cyanobacterial bloom occurrence and cyanobacterial species growth in cultures, respectively, have been controlled by the availability of both inorganic nitrogen and phosphorus [18–21]. However, cyanobacteria taxa such as N_2-fixing (diazotrophic) and non-N_2 fixing (nondiazotrophic) species have a variety of mechanisms to compete for nitrogen. A strong relationship between the growth and toxin synthesis of diazotrophic and nondiazotrophic cyanobacteria and inorganic dissolved N in the medium has commonly been reported in the literature [22–25]. Efforts often focus on total nitrogen (TN) and there exist important gaps in the understanding of how N speciation (NO_3^- and NH_4^+) influences cyanobacterial blooms and cyanotoxin synthesis. For example, in lakes showing symptoms of N limitation during late summer, numerous studies reported that cyanobacteria such as *Microcystis*, a non-N_2 fixing genus, have been shown to become dominant by rapidly assimilating recycled ammonium [26–29]. Indeed, laboratory and *in situ* studies have shown that cyanobacteria appear to out-compete other algal species for reduced N forms such as ammonium and urea [26, 27, 30]. For example, Donald et al. [31] reported that fertilization of the lake Wascana (Canada) with ammonium increased total algal abundance about 350% and cyanobacterial biomass over 500%. In a recent study examining the effects of different forms of N (nitrate, ammonium, and urea) in Lake Erie (U.S.A.), Chaffin et al. [32] have shown that the ammonium enrichment resulted in greater cyanobacterial biovolume than in the nitrate and urea enrichments. While nitrate is generally the most abundant form of nitrogen in freshwater ecosystems, it is the least preferred form of nitrogen, since its uptake by cyanobacteria requires multiple steps of intracellular reduction to ammonia [32–34]. Hence, while nitrogen (N) plays a primary role in shaping the relative abundance of cyanoHABs in a freshwater ecosystem, phosphorus (P) likely acts and interacts to influence these populations as well. The ability of cells to store phosphorus as polyphosphates [35] allows them to double several times even in phosphorus-limiting conditions [36]. Phosphorus affinities are higher in cyanobacteria compared to eukaryotic algae [37]. The concentration of phosphorus around 0.03 mg L^{-1} is enough for the sufficient growth of the cyanobacteria [38]. Therefore, phosphorus is commonly considered to be the limiting nutrient in freshwater ecosystems, and high concentrations of this nutrient often correlate to the occurrence of cyanobacterial blooms worldwide [2, 6, 39–42]. In contrast, instead of considering the effects of N and P separately, numerous studies highlight the importance of the ratio of TN to TP (TN:TP) in determining cyanobacterial growth [42, 43]. For example, several studies in many freshwater bodies showed that when the ratio TN:TP decreased, a shift has been reported in phytoplankton assemblages toward cyanobacteria dominance [2, 12, 44].

Light intensity and quality are other important factors in phytoplankton growth. Phytoplankton can photosynthesize, using the pigments chlorophyll-a (Chl-a) and -b, therefore at a certain light intensity, depending on the species, the algae will be at maximum productivity. The pigments are also sensitive to specific wavelengths: blue and red light. Using two species as an example, *Aphanizomenon flos-aquae* is less light dependent than *Dolichospermum flos-aquae*, so in situations where there is less light, *Aphanizomenon flos-aquae* would be at maximum production. The cyanobacterial light harvesting mechanism is different from that of the eukaryotic algae and contains phycobiliproteins, which allows cyanobacteria to absorb light from a wide light spectrum [36]. In the fast-changing light environment, cyanobacteria have a photoadaptation mechanism, which reduces the number of harvesting mechanisms and turns the energy into the heat [36]. There is also a photoprotective mechanism that cyanobacteria use: energy dissipation mechanism [36]. They have a UV photoprotection mechanism as well: mycosporine-like amino acids (MAAs) and scytonemin that absorbs UV light and helps them to survive with high level of irradiance [45].

Water temperature leads to cyanobacterial bloom development and plays a critical role in buoyancy and assimilation of essential nutrients and synthesis of toxins [46–48]. For example, Kosten et al. [19] by examining 143 lakes along a latitudinal transect ranging from subarctic Europe to southern South America found that temperature and TN concentrations were the strongest explanatory variables for cyanobacterial biomass. This finding is also consistent with that reported by Beaulieu et al. [20] who examined the proliferation of cyanobacteria in 1147 lakes and reservoirs of different trophic status in the United States and showed that the best linear multiple regression model for predicting these events was based on TN and temperature of the lake water. Therefore, in terms of global climate change, it is obvious that the increase of water temperature will be observed, and cyanobacteria will have the prevalence in the growth rate compared to the other phytoplankton.

Turbidity is another factor that influences algal growth. The particulate matter in the water column affects light penetration and temperature of the water. An excess of sediments in the water would decrease the light penetration, which in turn may prevent large algal blooms. The sediments also aid in lowering the amount of temperature fluctuations in the water. A more consistent and possibly lower temperature would help prevent large algal blooms [49]. The higher the pH, the higher diversity of cyanobacteria can be found with a prevalence of nonfixers (*Microcystis* spp.), while N-fixers are more dominant at low pH [50]. The structure of the lake plays the accessory role in bloom formations. The depth of a water body, speed of flow, and presence of small coves make every water body unique and need additional attention.

Abiotic factors described above are not the only ecological factors influencing the occurrence and frequent dominance of cyanobacteria in the phytoplankton. Their widespread representation in freshwaters depends also on biotic factors such as buoyancy, allelopathic effects, and zooplankton grazing among others that will be examined in the next section.

2.2. Biotic factors

Cyanobacterial species have numerous physiological adaptations that permit them to exploit nutrients and light differentially. Some species belonging to the genera *Microcystis*, *Anabaena* (renamed *Dolichospermum*), *Planktothrix*, *Aphanizomenon*, and *Cylindrospermopsis* among others possess gas vesicles that provide them buoyancy and vertical movement through the water column and can therefore effectively dominate other pelagic planktonic algae for available sunlight and nutrients [51]. This physiological capacity confers a substantial ecological advantage to these species, as they can congregate at a dense mass in the water column of stratified lakes and move up and down in the water column to maximize photosynthesis in the surface layers where there is more photosynthetically active radiation and to take up nutrients in dark deeper layers where the concentration of nutrients is higher. In addition, the ability of these genera to fix and assimilate dissolved nitrogen gas when the external concentrations of dissolved nitrate and ammonia fall to low levels is a supplementary biotic factor that offers them an ecological advantage over other phytoplanktonic species [52, 53]. Moreover, the resistance of the larger, gas-vesicle colony-forming cyanobacteria such as *Microcystis* to sinking loss or consumption by grazers can provide a significant advantage when this factor operates against small, nonmotile unicellular phytoplankton [54–56]. In addition, some species of cyanobacteria produce allelopathic substances that prevent the growth of submerged vegetation and other algae [57, 58], as well as increased resistance to predation by zooplankton, reducing the diversity of grazing species, and therefore the formation of blooms [59]. Besides zooplankton grazing, fish activities, benthic fauna, bacteria relationship, and viral lyses are considered as supplementary biotic factors that control algal blooms [7].

3. Mathematical modeling: a necessary approach for studying cyanoHAB proliferation

3.1. General context

As we notice from previous parts, relationships between the bloom patterns and involved factors are highly complex, therefore appropriate prognostic techniques to forecast blooms and evaluate their spatio-temporal evolution are indispensable. However, due to the complicity and nonlinearity of the phenomenon, none of the research on predictive approaches seems accurate and none has performed well to date. Moreover, no existing research could help to identify the very important factor: thresholds of blooms under the environmental conditions. This part of the chapter will review some mainstreams of mathematical models used in the bloom prediction.

3.2. Deterministic versus probabilistic

With the development of super-powerful computers and computational techniques, many mathematical models for predicting the algal growth have been developed in recent years. There are two main families of mathematical models which are commonly used: deterministic and probabilistic (or stochastic). The deterministic approach can be chosen when the nature of problem and dataset have well determined, repeatable, and fixed outputs for the same inputs. This means they have a precise *cause-effect* relationship. Conversely, a stochastic approach is preferable when a system has some inherent *randomness* and we must estimate possible outcomes with their *occurrence probability* to define its behavior. Stochastic models are fundamentally built based on the randomness and uncertainty of the nature of bloom pattern, reflected through a large amount of field data. These data are indispensable for the modeler to validate and test the precision and accuracy of their models. Among the category of stochastic models, machine learning techniques provide many powerful tools to solve some relevant difficulties in predicting HAB. Machine learning techniques were developed for quantitative finance, enabling researchers to tap huge datasets. There are many publications in recent years in which diverse Supervised Machine Learning (SML) models have been applied to solve a wide range of problems, including the Artificial Neural Network (ANN) and Support Vector Machine (SVM). ANN is a SML approach widely used to predict the algal abundance [60–64] while SVM is used much less in algal research [65–67]. Some studies used genetic algorithms (GA) to create prediction model [68, 69]. The basic concept of these models lies in the combined effects of a set of explanatory variables (Xi) on one or some target variables (Yi), and then classification or regression decision depending on the nature of Yi. The variables Yi (outputs of the study) in most of the models target the pigment Chl-a, which stands for the growth of algal communities, biomass (algal abundance) quantity, and the number of algal cell counts.

Wei et al. [61] suggested a model to predict the timing and magnitude of four different types of algae: *Microcystis*, *Oscillatoria*, *Synedra*, and *Phormidium*. Algal blooms responded to the orthogonal combinations of water temperature, light penetration, dissolved oxygen, chemical oxygen demand, total nitrogen (TN), total phosphorus (TP), zooplankton, and pH value. This study used backpropagation ANN; data in this study were collected monthly during 15 years from 1982 to 1996 and these data were divided randomly into two sets: training dataset and testing dataset. This study also analyzed the sensitivity of the model in which pH played a key role in the blooming of algae and all four types of algae are more sensitive to TP than TN.

Another study conducted by Wilson and Recknagel [60] used feedforward ANN to predict the bloom of algae in Australia. They suggested a regression model between four inputs (phosphorus, nitrogen, underwater light, and water temperature) and one main output (biomass) was designed. In 30-day-ahead model, beside algae biomass, they added the second output chlorophyll-a and used time delay neuron network structure where inputs are one-time step (e.g., 30 days) in the past relative to output variables. Fernández et al. [70] suggested a model to predict the presence of cyanotoxins in fresh water in Spain. A group of six inputs consisting of both biological and chemical factors is used and the output is the presence of cyanotoxins ($\mu g\ L^{-1}$). The most significant aspect of this model is the product of the concentration of *M. aeruginosa* and *W. naegeliana*, followed by turbidity, total phosphorus, alkalinity, and water temperature. This model used generic algorithm (GA) and multivariate adaptive regression spline (MARS) techniques in which 10-fold cross validation was used to train and validate the model. Park et al. [62] developed an early-warning model for freshwater algal bloom based on Chl-a concentration using ANN and SVM. These authors used the weekly data in 7-year period (from 2006 to 2012) to design a 7-day interval prediction model for two lakes in Korea. Five water quality parameters including Chl-a, orthophosphate as phosphorus (PO_4-P), ammonium nitrogen (NH_3-N), nitrate nitrogen (NO_3-N), water temperature, and two meteorological data (solar radiation and wind speed) are inserted as inputs and output for ANN and SVM models.

Recently, Nelson et al. [71] used the Random Forest algorithm to characterize and quantify relationships between 10 different conditions and five dominant cyanobacteria genera. All explanatory variables were lagged by 1-month step to reflect the division rates of cyanobacteria in natural environments. Outputs are the biomass values of five different types of cyanobacteria genera.

As approaches using the probabilistic models are limited due to their complex concepts and high randomness levels, especially due to the needs of a large amount of data to validate them, which use various factors that influence cyanobacterial growth, the deterministic strategies will allow the evaluation of the risks associated with cyanobacteria in the context of "less data needed" and moreover, many physical parameters could be incorporated in coupling with biochemical factors. Various deterministic approaches [72–78] have been used in the understanding of the distribution of cyanobacteria. The Lagrangian deterministic model follows the cyanobacterial colony in the water column so that a mathematical model can be created to describe bloom density. The Kromkamp and Walsby [72] model is only used to estimate settling velocities and the Visser et al. [76] model is an improved model, which incorporates the irradiance-response curve of density change and proposed an equation that describes the rate of density change in the dark. The Lagrangian approach is used for studying movement of cells in a laboratory setting, but an Eulerian approach enables exploration of full-scale spatial distribution of cells at specific times. Bruggeman and Bolding [79] built a framework called the Fortran-based framework for aquatic biogeochemical models where the biochemical model was connected to a physical model. Then a self-contained complex biological model was combined with a hydrodynamic model by the Fortran-based framework for aquatic biogeochemical model. This model was used to calibrate physiological parameters for the phytoplankton. Recently, the work of Ndong et al. [80] has shown a sophisticated 2D Eulerian frame model to evaluate the phototactic behavior effect of cyanobacteria, as well as the effects of light and wind on the distribution of cyanobacteria and estimate coupled effects of biological and physical factors on cyanobacteria.

The new tendency of research based on the deterministic approach is using remote sensing data or satellite imagery in the detection of the spatiotemporal patterns of blooms and explains how they change under the environmental conditions. The issue with this imagery is that the movement patterns of cells in the water column may be missed. The response to light intensity, nutrient levels, and temperature also needs to be considered, which means that numerical data along with imagery are required to complete the data. Agent-based models have been used to observe the 2D and 3D transport trajectories of cyanobacteria. These models are coupled with an Eulerian model, which allows the cyanobacteria to drift in the model [81–83].

3.3. Future perspectives

The overall and common goal of all models was to attempt to explain the risks of algal/cyanobacterial blooms and to study their evolution under environmental conditions leading to the improvement or decision process used to monitor cyanobacteria. However, as previously mentioned, almost all existing models have focused on the target variables such as Chl-a concentration development, cell count numbers of taxa or genera, biomass, etc.; from them, authors could conclude about the bloom situation. There are therefore two main directions of modeling among many others that should be developed: (1) determination of biophysical threshold for blooms and (2) quantifying and modeling the toxin concentration released by toxic species.

Remote sensing data combined with machine learning algorithm are also an encouraging perspective. But one of the potential pitfalls for machine learning strategies is the extremely low signal-to-noise ratio. Machine learning algorithms will always identify a pattern, even if there is none. In other words, the algorithms can view flukes as patterns and hence are likely to identify false strategies. Every model regardless of what category it belongs to can have its weak and strong points and need a serious validation step to be universally applicable, useful, and accurate.

4. Dominant taxa found in cyanoHABs in freshwater

Although cyanobacterial blooms are a worldwide phenomenon, there are differences in typical genera found in temperate and tropical regions (**Table 1**). *Microcystis* was the most frequently occurring bloom genus throughout the world, while *Cylindrospermopsis* and *Dolichospermum* (Syn. *Anabaena*) blooms occurred in various tropical areas such as Australia, America, and Africa.

Region	Dominant species	References
Africa	*Microcystis flos-aquae, M. wesenbergii, Oscillatoria* sp., *Dolichospermum* sp., *Lingbya* sp., *Anabaenopsis* sp.	[3, 48, 84, 85]
Western Asia	*Planktothrix rubescens, M. aeruginosa, Nodularia spumigena, Aphanizomenon ovalisporum, P. agardhii, Synechocystis* sp., *Dolichospermum* sp.	[84, 86]
Southern Asia	*Dolichospermum* sp., *Aphanizomenon* sp., *Microcystis* sp., *Cylindrospermopsis* sp., *Planktothrix* sp.	[84, 86]
Eastern Asia	*Dolichospermum* sp., *Microcystis* sp., *Aphanizomenon* sp., *Merismopedia* sp., *Cylindrospermopsis* sp., *Nostoc* sp., *Planktothrix* sp.	[7, 84, 85]
Oceania	*D. planctonicum, D. circinale, Aphanizomenon tenuicaulis, C. raciborskii, A. ovalisporum, A. issatschenkoi, P. rubescens, Kamptonema formosum, M. aeruginosa, M. panniformis*	[48, 84–86]
South and Central America	*M. aeruginosa, Cylindrospermopsis* sp., *Dolichospermum* sp., *Nodularia* sp., *Lingbya* sp.	[7, 48, 84, 85]
North America	*M. aeruginosa, M. viridis, M. wesenbergii, Aphanizomenon schindleri, D. flos-aquae, D. planctonicum, D. circinale, D. lemmermani, D. smithii, D. viquiera, C. raciborskii, P. rubescens, Lyngbya majuscula, L. wollei, Phormidium* sp., *Woronichinia naegeliana*	[7, 87–89]
Europe	*Microcystis* sp., *Dolichospermum* sp., *Aphanizomenon* sp., *Planktothrix* sp., *Nodularia* sp., *Cylindrospermopsis* sp., *Phormidium* sp., *Anabaenopsis* sp., *Gloeotrichia* sp.	[7, 84, 86, 90]

Table 1. Dominant cyanobacterial taxa recorded worldwide.

5. Negative outcomes from cyanoHABs

5.1. Cyanobacterial toxins and their environmental concentrations

Cyanotoxins are classified according to their mode of action into three families: neurotoxins (nervous system), hepatotoxins (liver), and dermotoxins (skin) [4, 17, 91]. Blooms formed by cyanobacteria producing hepatotoxins (microcystins and cylindrospermopsin) are more widespread than neurotoxic blooms [4, 92–95] and therefore, they are considered priority for biomonitoring, especially in drinking and recreational waters.

Cyanotoxins are intracellular toxins that are released into water only during cellular senescence or death and lysis or through water treatment processes such as application of algaecide [96, 97]. Therefore, total concentrations (intracellular plus extracellular) of microcystins, the most common cyanotoxins, vary from trace to several milligrams per liter [91, 98]. For example, very high concentrations have been reported up to 8428 μg L^{-1} in Southwest wetlands, Australia [99], 19,500 μg L^{-1} in Lake Suwa, Japan [41], 23,718 μg L^{-1} in Dam Nhanganzwane, South Africa [100], 29,200 μg L^{-1} in Lake Oubeira, Algeria [101], or 36,500 μg L^{-1} in Lake Horowhenua, New Zealand [102]. Messineo et al. [103] reported that in several Italian lakes, concentrations of total cylindrospermopsin varied from nondetectable values up to 126 μg L^{-1}. However, neurotoxins are less common in the freshwater ecosystems. For example, Rapala et al. [104] reported up to 1070 μg L^{-1} of saxitoxin in Finnish lakes. Anatoxin-a was detected in two shallow reservoirs (Konstantynów and Kraśnik) in Poland at concentrations ranging from 0.03 to 43.6 μg L^{-1} during a bloom of *D. flos-aquae* [105]. Recently, Roy-Lachapelle et al. [106] reported that the concentrations of the BMAA in 12 different lake waters in Canada ranged between 0.009 and 0.3 μg L^{-1}.

5.2. Ecotoxicological effects of cyanotoxins

Cyanotoxins such as hepatotoxins and neurotoxins target in humans and animals the liver and nervous system, respectively, but they often have important side effects too. When present in freshwater ecosystems, they may also affect organisms at different trophic levels, especially those having identical or similar target organs, tissues, or cells.

5.2.1. *Acute effects*

The occurrence of cyanoHABs in aquatic ecosystems is often associated with fish mortality (**Figure 1**). In addition, terrestrial organisms such as livestock, dogs, and birds that are associated with these freshwater ecosystems in which cyanoHABs occur may also be at risk of cyanotoxins exposure from preying on toxic aquatic prey and/or drinking contaminated water. For example, Georges Francis was the first in 1878 to implicate cyanobacteria in the poisoning of farm animals, in Alexandrina Lake, Milang, Southern Australia [107]. Since then, a significant number of cases of animal poisonings attributable to cyanotoxins have been documented worldwide [108–111]. Fish and invertebrates which are exposed over their entire life cycle to cyanotoxins are the most aquatic organisms affected, followed by birds, livestock and poultry, and dogs [111]. Acute ecotoxicity data of cyanotoxins were compiled by several studies [112–114]. The most documented cyanotoxin effects are those on microcystins due to their occurrence at high concentrations up to 28 mg L^{-1} and the dominance of cyanobacterial species producing them [91]. In addition, depending on their mechanism of action as potent and specific inhibitors of protein phosphatases and inducer of oxidative stress, microcystins can affect a range of invertebrate and vertebrate organisms [109, 114, 115]. Therefore, they cause changes in the trophic levels and adverse impacts on the functioning of freshwater ecosystems. This begins with the zooplankton community, which has its composition changed, especially by the mortality of certain species resulting therefore in the reduction of their diversity [59, 109, 116, 117]. For example, the copepod *Diaptomus birgei* was the most sensitive to microcystins with a lethal concentration (LC_{50}) at 48 h of 0.45 to 1.0 μg mL^{-1} followed by the cladoceran *Daphnia pulex*, *D. hyaline*, and *D. pulicaria* with LC_{50} at 48 h of 9.6, 11.6, and 21.4 μg L^{-1}, respectively [109].

Figure 1. *Microcystis* sp. bloom associated with fish mortality (photo: N.Y. Benayache).

However, mollusks and decapods appeared to be relatively tolerant to microcystins [109, 111]. For example, the LC_{50} at 96 h for microcystin-LR equivalent in the decapod *Kalliapseudes schubartii* [118] and the crayfish *Procambarus clarkia* [119] is 1.58 and 0.567 mg L^{-1}, respectively. Similarly, bivalves bioaccumulate high concentrations of microcystins without symptoms of acute toxicity [120].

Mass mortalities of fish have also been attributed to microcystins [14, 111, 121]. However, some studies suggested that most of fish mortalities can also be attributed to hypoxic conditions resulting from bloom respiration and senescence and not only to cyanotoxins [122, 123]. Like bivalves, fish appear to be less sensitive to toxin's short-term exposure than zooplankton. For example, experimental investigations on the rainbow trout have been shown that this fish species appeared to be relatively tolerant to high concentrations of microcystin-LR and death was recorded only at 1000 μg kg^{-1} bw [124]. Several studies reported that the dose inducing the mortality of the half of the test population (LD_{50}) of microcystin-LR in fish ranges from 20 to 1500 μg kg^{-1} body weight [125].

For the other classes of alkaloid cyanotoxins such as cylindrospermopsin and neurotoxins (anatoxins and saxitoxins), there are few or no studies that have examined their acute toxicity on aquatic organisms. For example, Ferrão-Filho et al. [126] reported that the exposure of three cladoceran species (*Daphnia gessneri, D. pulex,* and *Moina micrura*) to a saxitoxin-producer strain (T3) of *Cylindrospermopsis raciborskii* at cell densities of 10^3 and 10^4 cells/mL for 24 h resulted in a complete paralysis of *D. pulex*; however, *D. gessneri* was not sensitive and *M. micrura* was intermediate in sensitivity. Osswald et al. [127] demonstrated that when common carp *Cyprinus carpio* larvae were exposed to a lyophilized suspension (10^7 cells/mL) of a strain of *Anabaena* sp. producing anatoxin-a, all fish died between 24 and 29 h.

5.2.2. Subchronic and chronic effects

Aquatic organisms are continuously exposed over long periods of time or even their entire life cycle to cyanotoxins; therefore, evaluation of chronic effects of these toxins is important for an accurate environmental risk assessment. Several studies have shown that aquatic organisms that are exposed in the long term to cyanotoxins through the diet may die or display impaired feeding, immunosuppression, increased susceptibility to disease, avoidance behavior, physiological dysfunction, abnormal development, and reduced growth and reproduction

[109, 125]. For example, chronic exposure of parent *Daphnia magna* to either microcystin-LR at 5 or 50 μg L^{-1}, or to cyanobacterial crude extract containing the same amount of total microcystins, resulted in the decrease of the survival of offspring or cessation of eggs and reduced number of neonates and deformations of neonates such as incomplete development of the antennae [128]. Moreover, several studies have shown that when embryos and larvae of different species of fish including chub (*Leuciscus cephalus*), common carp (*Cyprinus carpio*), loach (*Misgurun smizolepis*), rainbow trout (*Oncorhynchus mykiss*) and zebrafish (*Danio rerio*) were immersed in solutions of 0.5–50 μg microcystins/L for up to 30 days, it resulted in interferences with hatching, developmental defects, liver damage, and/or increased mortality [129–132]. In another chronic study, Ernst et al. [133] by investing the effect of a high microcystin concentration on eggs and larvae of whitefish (*Coregonus lavaretus*) exposed to blooms of *Planktothrix* sp. during winter 1998 and 2000 in a Lake Ammersee (Germany) hatchery reported malformations of eggs and disturbances of reproduction success, suggesting that the disappearance of some coregonid age groups observed in this lake may be a result of these development effects of microcystins. In a laboratory study, oral subchronic exposure of the common carp (mean body weight of 322 g) to *Microcystis* by feeding with bloom scum at a dose of 50 μg microcystins/kg body weight for 28 days resulted in inhibition of growth, severe damage in hepatocytes, and significant increase of some plasmatic enzyme activities such as alanine aminotransferase and aspartate aminotransferase [134].

5.2.3. *Ecotoxicity of cyanotoxin mixtures*

Aquatic organisms are most likely subject to acute, subchronic, and chronic impacts resulting from exposure to a mixture class of cyanotoxins and not to individual toxins. Cyanobacterial species producing different toxins such as hepatotoxins, neurotoxins, and dermotoxins have been shown to coexist in blooms [91], therefore making the exposure to toxin mixtures a plausible scenario. To investigate this scenario with considering the possible synergistic toxicity of complex matrices, Esterhuizen-Londt et al. [135] tested the effects of two artificial toxin mixtures containing cyanobacterial hepatotoxins (microcystin-LR, -YR, and -RR), cyanobacterial hepatotoxins (microcystin-LR and cylindrospermopsin), and the neurotoxin β-N-methylamino-L-alanine hydrochloride, respectively, versus a crude cyanobacterial bloom extract (dominated by *Microcystis aeruginosa* with minor proportions of *Anabaena* sp. and *Oscillatoria* sp.) on the oxidative status of *Daphnia pulex*. The results showed that the cyanobacterial extract elicited higher oxidative stress response on *D. pulex* compared to exposure with the two artificial toxin mixtures. According to these studies, authors suggested that other unidentified compounds present in the cyanobacterial extract with synergistic effects may enhance the toxic effects. In fact, previous studies found stronger developmental effects of cyanobacterial extracts containing microcystins [136, 137] on the African clawed frog *Xenopus laevis* embryos and anatoxin-a [138] on common carp *Cyprinus carpio* larvae than their respective purified toxins.

In addition, in natural environments, cyanotoxins could interact with other anthropogenic micropollutants present in aquatic ecosystems and therefore, could attenuate or potentiate their adverse effects on aquatic organisms. For example, combined influence of microcystin-LR and a pesticide, carbaryl, was investigated on *Daphnia pulicaria* [139]. The results showed that the interaction between carbaryl and microcystins was highly significant and the two chemicals in a combinatorial exposure induced synergistic effects with frequent premature offspring delivery with body deformations including dented carapax or undeveloped heart. Furthermore, Cazenave et al. [140] observed less pronounced teratological effects within 24 h as well as nonsignificant increase in the activity of glutathione S-transferase (GST) in embryos of zebrafish (*Danio rerio*) exposed to either microcystin-LF or microcystin-RR in combination with natural organic matters compared to embryos exposed to pure toxins.

5.3. Bioaccumulation of cyanotoxins in food web and impacts on animal and human health

Zooplankton have been clearly identified as the best bioaccumulator of cyanotoxins and may transfer them to higher trophic levels in the aquatic food web [141–144]. Mollusks have also been shown to accumulate high concentrations of cyanotoxins with hepatopancreas being the organ presenting the highest concentrations followed by the intestines [115, 145]. As with invertebrates, fish can also accumulate high concentrations of microcystins but on average 3.5 times lower in planktivorous fish than in zooplankton [115]. For example, the highest concentrations of microcystins were found in the liver of the planktivorous fish *Osmerus eperlanus* reaching up to 874 μg microcystins/g dry weight [146]. However, in another planktivorous fish such as the silver carp *Hypophthalmichthys molitrix*, the highest concentrations of microcystins were found in the intestines reaching up to 137 $\mu g\ g^{-1}$ DW [147]. Carnivorous fish, meanwhile, accumulate less microcystins with the maximum concentration up to 51 $\mu g\ g^{-1}$ DW measured, for example, in the liver of perch *Perca fluviatilis* [146]. Overall, carnivorous fish, as superior predators, had lower mean microcystin content than planktivorous and omnivorous fish, suggesting transfer and bioaccumulation of microcystins, however without biomagnification in the food chain. In contrast, fish may act as an efficient vector of cyanotoxins to upper trophic levels such as birds and humans. In fact, numerous bird deaths have been reported in which most deaths are associated with the consumption of toxic prey, for example, fish or mollusks that have consumed or otherwise bioaccumulated cyanobacterial toxins [109, 110, 148, 149]. For humans, Chen et al. [150] confirmed for the first time the presence of microcystins in serum samples (average 0.39 ng/ml) of fishermen at Lake Chaohu, China. According this study, daily intake by the fishermen was estimated to be in the range of 2.2–3.9 μg MC-LR equivalent, whereas the provisional World Health Organization tolerable daily intake (TDI) for daily lifetime exposure is 0.04 $\mu g\ kg^{-1}$ or 2–3 μg per person. However, as has been described previously, the different species of fish accumulate microcystins mainly in the intestine and liver/hepatopancreas; this poses no risk to human health if these organs are taken from animals before consumption.

6. Conclusions

The chapter has sketched a general overview on cyanoHABs which recently become a real worrisome issue at the global scale due to their effects on water resources and animal and human health. They will cause ongoing issues as they will certainly reoccur over and over the coming years, especially under the promoting factors of climate changes and global warming effects, as much as the abuse of all watersheds due to anthropogenic actions.

Different research studies around the world have highlighted the complex relationships between cyanobacterial growth and environmental factors. The cyanoHAB dominance can result from a variety of interactions among biotic and abiotic components. The presence of toxic cyanobacteria can influence the human society at all scales, such as direct effects on drinking and recreational water resources, as well as the transfer of their toxins to higher trophic levels, resulting in fish kills and threats to all animal and human health.

Keeping our water resources clean, healthy, and safe for current and next generations becomes, therefore, a big challenge for our planet. The task of monitoring and managing cyanobacterial blooms and their negative outcomes including toxins released is a pressing concern for all. The chapter hence will serve to increase awareness of common challenges and existing capacities as well as lay the foundation for ongoing discussion and research on various subjects related to CyanoHABs that will be needed for effective management for the years to come.

Acknowledgements

NYB was granted by a fellowship from the Algerian-French PROFAS B + program. TNQ acknowledges the Natural Science and Engineering Research Council of Canada via Discovery Grant NSERC RGPIN 03796.

Conflict of interest

The authors wish to confirm that there are no known conflicts of interest associated with this chapter and there has been no significant financial support for this work that could have influenced its outcome. We confirm that the manuscript has been read and approved by all named authors and that there are no other persons who satisfied the criteria for authorship but are not listed. We further confirm that the order of authors listed in the manuscript has been approved by all of us.

Author details

Naila-Yasmine Benayache[1,3], Tri Nguyen-Quang[2], Kateryna Hushchyna[2], Kayla McLellan[2], Fatima-Zohra Afri-Mehennaoui[1] and Noureddine Bouaïcha[3]*

*Address all correspondence to: noureddine.bouaicha@u.psud.fr

1 Laboratory of Biology and Environment, University Frères Mentouri (Constantine 1), Constantine, Algeria

2 Biofluids and Biosystems Modeling Lab (BBML), Faculty of Agriculture, Dalhousie University, Truro, Nova Scotia, Canada

3 Laboratory Ecology, Systematic and Evolution, UMR 8079 Univ. Paris-Sud, CNRS, AgroParisTech, University Paris-Saclay, Orsay, France

References

[1] Steffen MM, Belisle BS, Watson SB, Boyer GL, Wilhelm SW. Status, causes and controls of cyanobacterial blooms in Lake Erie. Journal of Great Lakes Research. 2014;**40**:215-225. DOI: 10.1016/j.jglr.2013.12.012

[2] Gobler CJ, Burkholder JM, Davis TW, Harke MJ, Johengen T, Stow CA, et al. The dual role of nitrogen supply in controlling the growth and toxicity of cyanobacterial blooms. Harmful Algae. 2016;**54**:87-97. DOI: 10.1016/j.hal.2016.01.010

[3] Ndlela LL, Oberholster PJ, Van Wyk JH, Cheng PH. An overview of cyanobacterial bloom occurrences and research in Africa over the last decade. Harmful Algae. 2016;**60**:11-26. DOI: 10.1016/j.hal.2016.10.001

[4] Miller TR, Weirich C, Bartlett S. Cyanobacterial toxins of the laurentian great lakes, their toxicological effects, and numerical limits in drinking water. Marine Drugs. 2017;**15**:160. DOI: 10.3390/md15060160

[5] Paerl HW, Huisman J. Climate change: A catalyst for global expansion of harmful cyanobacterial blooms. Environmental Microbiology Reports. 2009;**1**:27-37. DOI: 10.1111/j.1758-2229.2008.00004.x

[6] O'Neil JM, Davis TW, Burford MA, Gobler CJ. The rise of harmful cyanobacteria blooms: The potential roles of eutrophication and climate change. Harmful Algae. 2012;**14**: 313-334. DOI: 10.1016/j.hal.2011.10.027

[7] Paerl HW, Otten TG. Harmful cyanobacterial blooms: Causes, consequences, and controls. Microbial Ecology. 2013;**65**:995-1010. DOI: 10.1007/s00248-012-0159-y

[8] Gehringer MM, Wannicke N. Climate change and regulation of hepatotoxin production in Cyanobacteria. FEMS Microbiology Ecology. 2014;**88**:1-25. DOI: 10.1111/1574-6941.12291

[9] Botana LM. Toxicological perspective on climate change: Aquatic toxins. Chemical Research in Toxicology. 2016;**29**:619-625. DOI: 10.1021/acs.chemrestox.6b00020

[10] Bullerjahn GS, McKay RM, Davis TW, Baker DB, Boyer GL, D'Anglada LV, et al. Global solutions to regional problems: Collecting global expertise to address the problem of harmful cyanobacterial blooms. A Lake Erie case study. Harmful Algae. 2016;**54**:223-238. DOI: 10.1016/j.hal.2016.01.003

[11] Paerl HW. Nuisance phytoplankton blooms in coastal, estuarine, and inland waters1. Limnology and Oceanography. 1988;**33**:823-843. DOI: 10.4319/lo.1988.33.4part2.0823

[12] Watson SB, McCauley E, Downing JA. Patterns in phytoplankton taxonomic composition across temperate lakes of differing nutrient status. Limnology and Oceanography. 1997;**42**:487-495. DOI: 10.4319/lo.1997.42.3.0487

[13] Funari E, Testai E. Human health risk assessment related to cyanotoxins exposure. Critical Reviews in Toxicology. 2008;**38**:97-125. DOI: 10.1080/10408440701749454

[14] Hilborn E, Beasley V. One health and cyanobacteria in freshwater systems: Animal illnesses and deaths are sentinel events for human health risks. Toxins. 2015;**7**:1374-1395. DOI: 10.3390/toxins7041374

[15] Carmichael WW, Boyer GL. Health impacts from cyanobacteria harmful algae blooms: Implications for the North American Great Lakes. Harmful Algae. 2016;**54**:194-212. DOI: 10.1016/j.hal.2016.02.002

[16] Bouaïcha N, Corbel S. Cyanobacterial toxins emerging contaminants in soils: A review of sources, fate and impacts on ecosystems, plants and animal and human health. In: Larramendy ML, Soloneski S, editors. Soil Contam. Curr. Consequences Furth. Solut. Rijeka: InTech; 2016. DOI: 10.5772/64940

[17] Buratti FM, Manganelli M, Vichi S, Stefanelli M, Scardala S, Testai E, et al. Cyanotoxins: Producing organisms, occurrence, toxicity, mechanism of action and human health toxicological risk evaluation. Archives of Toxicology. 2017;**91**:1049-1130. DOI: 10.1007/s00204-016-1913-6

[18] Dolman AM, Rücker J, Pick FR, Fastner J, Rohrlack T, Mischke U, et al. Cyanobacteria and cyanotoxins: The influence of nitrogen versus phosphorus. PLoS One. 2012;**7**:e38757. DOI: 10.1371/journal.pone.0038757

[19] Kosten S, Huszar VLM, Bécares E, Costa LS, van Donk E, Hansson L-A, et al. Warmer climates boost cyanobacterial dominance in shallow lakes. Global Change Biology. 2012;**18**:118-126. DOI: 10.1111/j.1365-2486.2011.02488.x

[20] Beaulieu M, Pick F, Gregory-Eaves I. Nutrients and water temperature are significant predictors of cyanobacterial biomass in a 1147 lakes data set. Limnology and Oceanography. 2013;**58**:1736-1746. DOI: 10.4319/lo.2013.58.5.1736

[21] Yuan LL, Pollard AI, Pather S, Oliver JL, D'Anglada L. Managing microcystin: Identifying national-scale thresholds for total nitrogen and chlorophyll a. Freshwater Biology. 2014;**59**:1970-1981. DOI: 10.1111/fwb.12400

[22] Orr PT, Jones GJ. Relationship between microcystin production and cell division rates in nitrogen-limited Microcystis aeruginosa cultures. Limnology and Oceanography. 1998;**43**:1604-1614. DOI: 10.4319/lo.1998.43.7.1604

[23] Vezie C, Rapala J, Vaitomaa J, Seitsonen J, Sivonen K. Effect of nitrogen and phosphorus on growth of toxic and nontoxic microcystis strains and on intracellular microcystin concentrations. Microbial Ecology. 2002;**43**:443-454. DOI: 10.1007/s00248-001-0041-9

[24] Davis TW, Harke MJ, Marcoval MA, Goleski J, Orano-Dawson C, Berry DL, et al. Effects of nitrogenous compounds and phosphorus on the growth of toxic and non-toxic strains of microcystis during cyanobacterial blooms. Aquatic Microbial Ecology. 2010;**61**:149-162. DOI: 10.3354/ame01445

[25] Monchamp M-E, Pick FR, Beisner BE, Maranger R. Nitrogen forms influence microcystin concentration and composition via changes in cyanobacterial community structure. PLoS One. 2014;**9**:e85573. DOI: 10.1371/journal.pone.0085573

[26] Ferber LR, Levine SN, Lini A, Livingston GP. Do cyanobacteria dominate in eutrophic lakes because they fix atmospheric nitrogen? Freshwater Biology. 2004;**49**:690-708. DOI: 10.1111/j.1365-2427.2004.01218.x

[27] McCarthy MJ, James RT, Chen Y, East TL, Gardner WS. Nutrient ratios and phytoplankton community structure in the large, shallow, eutrophic, subtropical Lakes Okeechobee (Florida, USA) and Taihu (China). Limnology. 2009;**10**:215-227. DOI: 10.1007/s10201-009-0277-5

[28] Chaffin JD, Bridgeman TB, Heckathorn SA, Mishra S. Assessment of microcystis growth rate potential and nutrient status across a trophic gradient in western Lake Erie. Journal of Great Lakes Research. 2011;**37**:92-100. DOI: 10.1016/j.jglr.2010.11.016

[29] Glibert PM, Wilkerson FP, Dugdale RC, Raven JA, Dupont CL, Leavitt PR, et al. Pluses and minuses of ammonium and nitrate uptake and assimilation by phytoplankton and implications for productivity and community composition, with emphasis on nitrogen-enriched conditions. Limnology and Oceanography. 2016;**61**:165-197. DOI: 10.1002/lno.10203

[30] Blomqvist P, Pettersson A, Hyenstrand P. Ammonium-nitrogen-Akey regulatory factor causing dominance of non-nitrogen-fixing cyanobacteria in aquatic systems. Archiv für Hydrobiologie. 1994;**132**:141-164

[31] Donald DB, Bogard MJ, Finlay K, Bunting L, Leavitt PR. Phytoplankton-specific response to enrichment of phosphorus-rich surface waters with ammonium, nitrate, and urea. PLoS One. 2013;**8**:e53277. DOI: 10.1371/journal.pone.0053277

[32] Chaffin JD, Davis TW, Smith DJ, Baer MM, Dick GJ. Interactions between nitrogen form, loading rate, and light intensity on Microcystis and Planktothrix growth and microcystin production. Harmful Algae. 2018;**73**:84-97. DOI: 10.1016/j.hal.2018.02.001

[33] Zhang H, Culver DA, Boegman L. A two-dimensional ecological model of Lake Erie: Application to estimate dreissenid impacts on large lake plankton populations. Ecological Modelling. 2008;**214**:219-241. DOI: 10.1016/j.ecolmodel.2008.02.005

[34] Harke MJ, Gobler CJ. Daily transcriptome changes reveal the role of nitrogen in controlling microcystin synthesis and nutrient transport in the toxic cyanobacterium, *Microcystis aeruginosa*. BMC Genomics. 2015;**16**:1068. DOI: 10.1186/s12864-015-2275-9

[35] Markou G, Vandamme D, Muylaert K. Microalgal and cyanobacterial cultivation: The supply of nutrients. Water Research. 2014;**65**:186-202. DOI: 10.1016/j.watres.2014.07.025

[36] Oliver RL, Hamilton DP, Brookes JD, Ganf GG. Physiology, blooms and prediction of planktonic cyanobacteria. In: Whitton BA, editor. Ecol. Cyanobacteria II Their Divers. Space Time. Dordrecht: Springer Netherlands; 2012. pp. 155-194. DOI: 10.1007/978-94-007-3855-3_6

[37] Molot LA, Watson SB, Creed IF, Trick CG, McCabe SK, Verschoor MJ, et al. A novel model for cyanobacteria bloom formation: The critical role of anoxia and ferrous iron. Freshwater Biology. 2014;**59**:1323-1340. DOI: 10.1111/fwb.12334

[38] Šejnohová L, Maršálek B. Microcystis. In: Whitton BA, editor. Ecology of Cyanobacteria II: Their Diversity in Space and Time. Netherlands, Dordrecht: Springer Science+Business Media B.V.; 2012. pp. 195-228. https://doi.org/10.1007/978-94-007-3855-3

[39] Xu H, Paerl HW, Qin B, Zhu G, Hall NS, Wu Y. Determining Critical nutrient thresholds needed to control harmful cyanobacterial blooms in Eutrophic Lake Taihu, China. Environmental Science & Technology. 2015;**49**:1051-1059. DOI: 10.1021/es503744q

[40] Huang L, Fang H, He G, Jiang H, Wang C. Effects of internal loading on phosphorus distribution in the Taihu Lake driven by wind waves and lake currents. Environmental Pollution. 2016;**219**:760-773. DOI: 10.1016/j.envpol.2016.07.049

[41] Harke MJ, Davis TW, Watson SB, Gobler CJ. Nutrient-controlled niche differentiation of western lake erie cyanobacterial populations revealed via metatranscriptomic surveys. Environmental Science & Technology. 2016;**50**:604-615. DOI: 10.1021/acs.est.5b03931

[42] Li J, Hansson L-A, Persson K. Nutrient control to prevent the occurrence of cyanobacterial blooms in a eutrophic lake in Southern Sweden, used for drinking water supply. Water. 2018;**10**:919. DOI: 10.3390/w10070919

[43] Jeppesen E, Kronvang B, Meerhoff M, Søndergaard M, Hansen KM, Andersen HE, et al. Climate change effects on runoff, catchment phosphorus loading and lake ecological state, and potential adaptations. Journal of Environmental Quality. 2009;**38**:1930-1941. DOI: 10.2134/jeq2008.0113

[44] Downing JA, McCauley E. The nitrogen:phosphorus relationship in lakes. Limnology and Oceanography. 1992;**37**:936-945. DOI: 10.4319/lo.1992.37.5.0936

[45] Paerl HW, Paul VJ. Climate change: Links to global expansion of harmful cyanobacteria. Water Research. 2012;**46**:1349-1363. DOI: 10.1016/j.watres.2011.08.002

[46] Davis TW, Berry DL, Boyer GL, Gobler CJ. The effects of temperature and nutrients on the growth and dynamics of toxic and non-toxic strains of Microcystis during cyanobacteria blooms. Harmful Algae. 2009;**8**:715-725. DOI: 10.1016/j.hal.2009.02.004

[47] Moss B. Allied attack: Climate change and eutrophication. Inland Waters. 2011;**1**:101-105. DOI: 10.5268/IW-1.2.359

[48] Mowe MAD, Mitrovic SM, Lim RP, Furey A, Yeo DCJ. Tropical cyanobacterial blooms: A review of prevalence, problem taxa, toxins and influencing environmental factors. Journal of Limnology. 2015;**74**:205-224. DOI: 10.4081/jlimnol.2014.1005

[49] Wang W-C. Effect of turbidity on algal. Growth. Circular 121. State of Illinois, Department of Registration and Education. 1974. pp. 1-12. ISWS-74-CIR121

[50] Whitton BA, Potts M. Introduction to the Cyanobacteria. In: Whitton BA, editor. Ecol. Cyanobacteria II Their Divers. Space Time. Dordrecht: Springer Netherlands; 2012. pp. 1-13. DOI: 10.1007/978-94-007-3855-3_1

[51] Oliver RL, Walsby AE. Direct evidence for the role of light mediated gas vesicle collapse in the buoyancy regulation of Anabaena flos-aquae (cyanobacteria). Limnology and Oceanography. 1984;**29**:879-886

[52] Dokulil MT, Teubner K. Cyanobacterial dominance in lakes. Hydrobiologia. 2000;**438**:1-12. DOI: 10.1023/A:1004155810302

[53] Paerl HW, Gardner WS, Havens KE, Joyner AR, McCarthy MJ, Newell SE, et al. Mitigating cyanobacterial harmful algal blooms in aquatic ecosystems impacted by climate change and anthropogenic nutrients. Harmful Algae. 2016;**54**:213-222. DOI: 10.1016/j.hal.2015.09.009

[54] Fulton RS, Paerl HW. Effects of the blue-green alga *Microcystis aeruginosa* on zooplankton competitive relations. Oecologia. 1988;**76**:383-389. DOI: 10.1007/BF00377033

[55] Lampert W. Laboratory studies on zooplankton-cyanobacteria interactions. New Zealand Journal of Marine and Freshwater Research. 1987;**21**:483-490. DOI: 10.1080/00288330.1987.9516244

[56] Wilson AE, Sarnelle O, Tillmanns AR. Effects of cyanobacterial toxicity and morphology on the population growth of freshwater zooplankton: Meta-analyses of laboratory experiments. Limnology and Oceanography. 2006;**51**:1915-1924. DOI: 10.4319/lo.2006.51.4.1915

[57] Figueiredo MAT, Nowak RD, Wright SJ. Gradient projection for sparse reconstruction: Application to compressed sensing and other inverse problems. IEEE Journal of Selected Topics in Signal Processing. 2007;**1**:586-597. DOI: 10.1109/JSTSP.2007.910281

[58] Antunes JT, Leão PN, Vasconcelos VM. Influence of biotic and abiotic factors on the allelopathic activity of the cyanobacterium *Cylindrospermopsis raciborskii* strain LEGE 99043. Microbial Ecology. 2012;**64**:584-592. DOI: 10.1007/s00248-012-0061-7

[59] Azevêdo DJS, Barbosa JEL, Porto DE, Gomes WIA, Molozzi J. Biotic or abiotic factors: Which has greater influence in determining the structure of rotifers in semi-arid reservoirs? Acta Limnologica Brasiliensia. 2015;**27**:60-77. DOI: 10.1590/S2179-975X2914

[60] Wilson H, Recknagel F. Towards a generic artificial neural network model for dynamic predictions of algal abundance in freshwater lakes. Ecological Modelling. 2001;**146**(1-3): 69-84

[61] Wei B, Sugiura N, Maekawa T. Use of artificial neural network in the prediction of algal blooms. Water Research. 2001;**35**(8):2022-2028

[62] Park Y, Cho KH, Park J, Cha SM, Kim JH. Development of early-warning protocol for predicting chlorophyll-a concentration using machine learning models in freshwater and estuarine reservoirs, Korea. Science of the Total Environment. 2015;**502**:31-41

[63] Lu F, Chen Z, Liu W, Shao H. Modeling chlorophyll-a concentrations using an artificial neural network for precisely eco-restoring lake basin. Ecological Engineering. 2016;**95**: 422-429

[64] Tian W, Liao Z, Zhang J. An optimization of artificial neural network model for predicting chlorophyll dynamics. Ecological Modelling. 2017;**364**:42-52

[65] Liu J, Zhang Y, Qian X. Modeling chlorophyll-a in Taihu Lake with machine learning models. In: Proceedings of 3rd International Conference on Bioinformatics and Biomedical Engineering (ICBBE). Beijing, China; 2009. pp. 1-6

[66] Liu Z, Wang X, Cui L, Lian X, Xu J. Research on water bloom prediction based on least squares support vector machine. In: WRI World Congress on Computer Science and Information Engineering. Vol. 5. Los Angeles, California, USA: IEEE; 2009. pp. 764-768

[67] Xie Z, Lou I, Ung WK, Mok KM. Freshwater algal bloom prediction by support vector machine in Macau storage reservoirs. Mathematical Problems in Engineering. 2012;**2012**:1-12. Article ID: 397473. https://doi.org/10.1155/2012/397473

[68] Bobbin J, Recknagel F. Inducing explanatory rules for the prediction of algal blooms by genetic algorithms. Environment International. 2001;**27**(2-3):237-242

[69] Sivapragasam C, Muttil N, Muthukumar S, Arun VM. Prediction of algal blooms using genetic programming. Marine Pollution Bulletin. 2010;**60**(10):1849-1855

[70] Fernández JA, Muñiz CD, Nieto PG, de Cos Juez FJ, Lasheras FS, Roqueñí MN. Forecasting the cyanotoxins presence in fresh waters: A new model based on genetic algorithms combined with the MARS technique. Ecological Engineering. 2013;**53**:68-78

[71] Nelson NG, Muñoz-Carpena R, Phlips EJ, Kaplan D, Sucsy P, Hendrickson J. Revealing biotic and abiotic controls of harmful algal blooms in a shallow subtropical lake through statistical machine learning. Environmental Science & Technology. 2018;**52**(6):3527-3535

[72] Kromkamp J, Walsby AE. A computer model of buoyancy and vertical migration in cyanobacteria. Journal of Plankton Research. 1990;**12**:161-183. DOI: 10.1093/plankt/12.1.161

[73] Webster IT. Effect of wind on the distribution of phytoplankton cells in lakes. Limnology and Oceanography. 1990;**35**:989-1001. DOI: 10.4319/lo.1990.35.5.0989

[74] Verhagen JHG. Modeling phytoplankton patchiness under the influence of wind-driven currents inlakes. Limnology and Oceanography. 1994;**39**:1551-1565. DOI: 10.4319/lo.1994.39.7.1551

[75] Webster IT, Hutchinson PA. Effect of wind on the distribution of phytoplankton cells in lakes revisited. Limnology and Oceanography. 1994;**39**:365-373. DOI: 10.4319/lo.1994.39.2.0365

[76] Visser PM, Passarge J, Mur LR. Modelling vertical migration of the cyanobacterium Microcystis. Hydrobiologia. 1997;**349**:99-109. DOI: 10.1023/A:1003001713560

[77] Wallace BB, Hamilton DP. Simulation of water-bloom formation in the cyanobacterium Microcystis aeruginosa. Journal of Plankton Research. 2000;**22**:1127-1138. DOI: 10.1093/plankt/22.6.1127

[78] Porat R, Teltsch B, Perelman A, Dubinsky Z. Diel buoyancy changes by the cyanobacterium *Aphanizomenon ovalisporum* from a shallow reservoir. Journal of Plankton Research. 2001;**23**:753-763. DOI: 10.1093/plankt/23.7.753

[79] Bruggeman J, Bolding K. A general framework for aquatic biogeochemical models. Environmental Modelling & Software. 2014;**61**:249-265. DOI: 10.1016/j.envsoft.2014.04.002

[80] Ndong M, Bird D, Nguyen Quang T, Kahawita R, Hamilton D, de Boutray ML, et al. A novel Eulerian approach for modelling cyanobacteria movement: Thin layer formation and recurrent risk to drinking water intakes. Water Research. 2017;**127**:191-203. DOI: 10.1016/j.watres.2017.10.021

[81] Feng T, Wang C, Wang P, Qian J, Wang X. How physiological and physical processes contribute to the phenology of cyanobacterial blooms in large shallow lakes: A new Euler-Lagrangian coupled model. Water Research. 2018;**140**:34-43. DOI: 10.1016/j.watres.2018.04.018

[82] Li Y, Zhang Q, Ye R, Yao J, Tan Z. 3D hydrodynamic investigation of thermal regime in a large river-lake-floodplain system (Poyang Lake, China). Journal of Hydrology. 2018;**567**:86-101. DOI: 10.1016/j.jhydrol.2018.10.007

[83] Pyo J, Ligaray M, Kwon Y, Ahn M-H, Kim K, Lee H, et al. High-spatial resolution monitoring of phycocyanin and chlorophyll-a using airborne hyperspectral imagery. Remote Sensing. 2018;**10**:1180. DOI: 10.3390/rs10081180

[84] Codd GA, Azevedo SMFO, Bagchi SN, Burch MD, Carmichael WW, Harding WR, et al. CYANONET a global network for cyanobacterial bloom and toxin risk management. Initial situation assessment and recommendations. IHP-VI Technical Document in Hydrology. 2005;**76**. UNESCO Working Series SC-2005/WS/55

[85] Blaha L, Babica P, Marsalek B. Toxins produced in cyanobacterial water blooms—Toxicity and risks. Interdisciplinary Toxicology. 2009;**2**(2):36-41. DOI: 10.2478/v10102-009-0006-2

[86] Meriluo J, Blaha L, Bojadzija G, Bormans M, Brient L, Codd GA, et al. Toxic cyanobacteria and cyanotoxins in European waters—Recent progress achieved through the CYANOCOST. Action and challenges for further research. Advances in Oceanography and Limnology. 2017;**8**(1):161-178. DOI: 10.4081/aiol.2017.6429*AR*

[87] Schanz F, Allen ED, Gorham PR. Bioassay of the seasonal ability of water from a eutrophic Alberta lake to promote selective growth of strains of *Anabaena flos-aquae*and other blue-green algae. Canadian Journal of Botany. 1979;**57**(21):2443-2451. DOI: 10.1139/b79-288

[88] Pick FR. Blooming algae: A Canadian perspective on the rise of toxic cyanobacteria. Canadian Journal of Fisheries and Aquatic Sciences. 2016;**73**:1149-1158

[89] Nguyen-Quang T, Fraser H, Hushchyna K, McLellan K, Sabir Q. Continuation of the systematic study (2nd year – 2017) of toxic algal blooms in McLaughlin and Irishtown Reservoirs to add to the water quality program of the City of Moncton—New Brunswick. Technical Report submitted to Government of New Brunswick, Environmental Trust Fund, Environmental and Local Government, Canada. 2018

[90] Svirčev Z, Simeunović J, Subakov-Simić G, Krstić S, Vidović M. Freshwater cyanobacterial blooms and cyanotoxin production in serbia in the past 25 years. Geographica Panonica. 2007;**11**:32-38

[91] Corbel S, Mougin C, Bouaïcha N. Cyanobacterial toxins: Modes of actions, fate in aquatic and soil ecosystems, phytotoxicity and bioaccumulation in agricultural crops. Chemosphere. 2014;**96**:1-15. DOI: 10.1016/j.chemosphere.2013.07.056

[92] Merel S, Walker D, Chicana R, Snyder S, Baurès E, Thomas O. State of knowledge and concerns on cyanobacterial blooms and cyanotoxins. Environment International. 2013;**59**:303-327. DOI: 10.1016/j.envint.2013.06.013

[93] Antunes JT, Leão PN, Vasconcelos VM. Cylindrospermopsis raciborskii: Review of the distribution, phylogeography, and ecophysiology of a global invasive species. Frontiers in Microbiology. 2015;**6**:1-12. DOI: 10.3389/fmicb.2015.00473

[94] Burford MA, Beardall J, Willis A, Orr PT, Magalhaes VF, Rangel LM, et al. Understanding the winning strategies used by the bloom-forming cyanobacterium *Cylindrospermopsis raciborskii*. Harmful Algae. 2016;**54**:44-53. DOI: 10.1016/j.hal.2015.10.012

[95] Preece EP, Hardy FJ, Moore BC, Bryan M. A review of microcystin detections in Estuarine and Marine waters: Environmental implications and human health risk. Harmful Algae. 2017;**61**:31-45. DOI: 10.1016/j.hal.2016.11.006

[96] Sivonen K, Jones G-J. Cyanobacterial toxins. In: Chorus I, Bartram J, editors. Toxiccyanobacteria in Water: A Guide to Their Public Health Consequences, Monitoring and Management. London and New York: World Health Organization, Taylor and Francis; 1999. pp. 41-111

[97] Saker ML, Griffiths DJ. The effect of temperature on growth and cylindrospermopsin content of seven isolates of *Cylindrospermopsis raciborskii* (Nostocales, Cyanophyceae) from water bodies in northern Australia. Phycologia. 2000;**39**:349-354. DOI: 10.2216/i0031-8884-39-4-349.1

[98] Pham T-L, Utsumi M. An overview of the accumulation of microcystins in aquatic ecosystems. Journal of Environmental Management. 2018;**213**:520-529. DOI: 10.1016/j.jenvman.2018.01.077

[99] Kemp A, John J. Microcystins associated with Microcystis dominated blooms in the Southwest wetlands, Western Australia. Environmental Toxicology. 2006;**21**:125-130. DOI: 10.1002/tox.20164

[100] Oberholster PJ, Myburgh JG, Govender D, Bengis R, Botha AM. Identification of toxigenic Microcystis strains after incidents of wild animal mortalities in the Kruger National Park, South Africa. Ecotoxicology and Environmental Safety. 2009;**72**:1177-1182

[101] Nasri A-B, Bouaıcha N, Fastner J. First report of a Microcystin-containing bloom of the cyanobacteria Microcystis spp. in Lake Oubeira, Eastern Algeria. Archives of Environmental Contamination and Toxicology. 2004;**46**:197-202. DOI: 10.1007/s00244-003-2283-7

[102] Wood SA, Holland PT, Stirling DJ, Briggs LR, Sprosen J, Ruck JG, et al. Survey of cyanotoxins in New Zealand water bodies between 2001 and 2004. New Zealand Journal of Marine and Freshwater Research. 2006;**40**:585-597. DOI: 10.1080/00288330.2006.9517447

[103] Messineo V, Bogialli S, Melchiorre S, Sechi N, Lugliè A, Casiddu P, et al. Cyanobacterial toxins in Italian freshwaters. Limnologica. 2009;**39**:95-106. DOI: 10.1016/j.limno.2008.09.001

[104] Rapala J, Berg KA, Lyra C, Niemi RM, Manz W, et al. *Paucibacter toxinivorans* gen. nov., sp. nov., a bacterium that degrades cyclic cyanobacterial hepatotoxins microcystins and nodularin. International Journal of Systematic and Evolutionary Microbiology. 2005;**55**:1563-1568. DOI: 10.1099/ijs.0.63599-0

[105] Pawlik-skowronska B, Toporowska M. Blooms of toxin-producing Cyanobacteria—A real threat in small dam reservoirs at the beginning of their operation. Oceanological and Hydrobiological Studies. 2011;**40**:30-37. DOI: 10.2478/s13545-011-0038-z

[106] Roy-Lachapelle A, Solliec M, Sauvé S. Determination of BMAA and three alkaloid cyanotoxins in lake water using dansyl chloride derivatization and high-resolution mass spectrometry. Analytical and Bioanalytical Chemistry. 2015;**407**:5487-5501. DOI: 10.1007/s00216-015-8722-2

[107] Francis G. Poisonous Australian lake. Nature. 1878;**18**:11

[108] Duy TN, Lam PKS, Shaw GR, Connell DW. Toxicology and risk assessment of freshwater cyanobacterial (Blue-Green Algal) toxins in water. In: Ware GW, editor. Rev. Environ. Contam. Toxicol. Contin. Residue Rev., New York, NY: Springer New York; 2000. pp. 113-185. DOI: 10.1007/978-1-4757-6429-1_3

[109] Landsberg JH. The effects of harmful algal blooms on aquatic organisms. Reviews in Fisheries Science. 2002;**10**:113-390. DOI: 10.1080/20026491051695

[110] Stewart I, Seawright AA, Shaw GR. Cyanobacterial poisoning in livestock, wild mammals and birds—An overview. In: Hudnell HK, editor. Cyanobacterial Harmful Algal Blooms State Sci. Res. Needs, New York, NY: Springer New York; 2008. pp. 613-637. DOI: 10.1007/978-0-387-75865-7_28

[111] Wood R. Acute animal and human poisonings from cyanotoxin exposure—A review of the literature. Environment International. 2016;**91**:276-282. DOI: 10.1016/j.envint.2016.02.026

[112] Wiegand C, Pflugmacher S. Ecotoxicological effects of selected cyanobacterial secondary metabolites a short review. Toxicology and Applied Pharmacology. 2005;**203**: 201-218. DOI: 10.1016/j.taap.2004.11.002

[113] Zanchett G, Oliveira-Filho E. Cyanobacteria and cyanotoxins: From impacts on aquatic ecosystems and human health to anticarcinogenic effects. Toxins. 2013;**5**:1896-1917. DOI: 10.3390/toxins5101896

[114] Bownik A. Harmful algae: Effects of cyanobacterial cyclic peptides on aquatic invertebrates—A short review. Toxicon. 2016;**124**:26-35. DOI: 10.1016/j.toxicon.2016.10.017

[115] Ferrão-Filho AS, Kozlowsky-Suzuki B. Cyanotoxins: Bioaccumulation and effects on aquatic animals. Marine Drugs. 2011;**9**:2729-2772. DOI: 10.3390/md9122729

[116] Vasconcelos JF, Barbosa JEL, Lira W, Azevedo SMFO. Microcystin bioaccumulation can cause potential mutagenic effects in farm fish. Egyptian Journal of Aquatic Research. 2013;**39**:185-192. DOI: 10.1016/j.ejar.2013.11.002

[117] Vilar MCP, de Araújo-Castro CMV, Moura ADN. Acute toxicity of Microcystis spp. (Cyanobacteria) bloom on *Moina minuta* (Cladocera) in a tropical reservoir, Northeastern

Brazil. Ecotoxicology and Environmental Contamination. 2014;**9**:93-98. DOI: 10.5132/eec.2014.01.012

[118] Montagnolli W, Zamboni A, Luvizotto-Santos R, Yunes JS. Acute effects of *Microcystis aeruginosa* from the Patos Lagoon Estuary, Southern Brazil, on the Microcrustacean *Kalliapseudes schubartii* (Crustacea: Tanaidacea). Archives of Environmental Contamination and Toxicology. 2004;**46**:463-469. DOI: 10.1007/s00244-003-2304-6

[119] An Z, Sun L, Wang P. Acute toxicity and accumulation of microcystin-leucine-arginine in the crayfish *Procambarus clarkii* (Girard, 1852). Crustaceana. 2015;**88**:397-404. DOI: 10.1163/15685403-00003424

[120] Amorim Á, Vasconcelos V. Dynamics of microcystins in the mussel Mytilus galloprovincialis. Toxicon. 1999;**37**:1041-1052. DOI: 10.1016/S0041-0101(98)00231-1

[121] de Oliveira Azavedo SMF, Carmouze J-P. Une mortalité de poissons dans une lagune tropicale (Brésil) durant une période de dominance de Cyanophyceae: Coïncidence ou consequence? Revue d'Hydrobiologie Tropicale. 1994;**27**:265-272

[122] Jewel MA, Affan MA, Khan S. Fish mortality due to cyanobacterial bloom in an aquaculture pond in Bangladesh. Pakistan Journal of Biological Sciences. 2003;**6**:1046-1050

[123] Ibelings BW, Havens KE. Cyanobacterial toxins: A qualitative meta–analysis of concentrations, dosage and effects in freshwater, estuarine and marine biota. In: Hudnell HK, editor. Cyanobacterial Harmful Algal Blooms State Sci. Res. Needs, New York, NY: Springer New York; 2008. pp. 675-732. DOI: 10.1007/978-0-387-75865-7_32

[124] Kotak BG, Zurawell RW, Prepas EE, Holmes CF. Microcystin-LR concentration in aquatic food web compartments from lakes of varying trophic status. Canadian Journal of Fisheries and Aquatic Sciences. 1996;**53**:1974-1985. DOI: 10.1139/f96-135

[125] Malbrouck C, Kestemont P. Effects of microcystins on fish. Environmental Toxicology and Chemistry. 2006;**25**:72. DOI: 10.1897/05-029R.1

[126] Ferrão-Filho AdS, da Costa SM, Ribeiro MGL, Azevedo SMFO. Effects of a saxitoxin-producer strain of *Cylindrospermopsis raciborskii* (cyanobacteria) on the swimming movements of cladocerans. Environmental Toxicology. 2008;**23**:161-168. DOI: 10.1002/tox.20320

[127] Osswald J, Rellán S, Gago A, Vasconcelos V. Toxicology and detection methods of the alkaloid neurotoxin produced by cyanobacteria, anatoxin-a. Environment International. 2007;**33**:1070-1089. DOI: 10.1016/j.envint.2007.06.003

[128] Dao TS, Do-Hong L-C, Wiegand C. Chronic effects of cyanobacterial toxins on Daphnia magna and their offspring. Toxicon. 2010;**55**:1244-1254. DOI: 10.1016/j.toxicon.2010.01.014

[129] Oberemm A, Fastner J, Steinberg CEW. Effects of microcystin-LR and cyanobacterial crude extracts on embryo-larval development of zebrafish (*Danio rerio*). Water Research. 1997;**31**:2918-2921. DOI: 10.1016/S0043-1354(97)00120-6

[130] Oberemm A, Becker J, Codd GA, Steinberg C. Effects of cyanobacterial toxins and aqueous crude extracts of cyanobacteria on the development of fish and amphibians.

Environmental Toxicology. 1999;**14**:77-88. DOI: 10.1002/(SICI)1522-7278(199902)14:1<77::AID-TOX11>3.0.CO;2-F

[131] Liu Y, Song L, Li X, Liu T. The toxic effects of microcystin-LR on embryo-larval and juvenile development of loach, *Misgurunsmizolepis gunthe*. Toxicon. 2002;**40**:395-399. DOI: 10.1016/S0041-0101(01)00173-8

[132] Palikova M, Navratil S, Marsalek B, Blaha L. Toxicity of crude extract ofcyanobacteria for embryos and larvae of carp (*Cyprinus carpio* L.). Acta Veterinaria Brunensis. 2003;**72**(3):437-443

[133] Ernst B, Hitzfeld B, Dietrich D. Presence of Planktothrix sp. and cyanobacterial toxins in Lake Ammersee, Germany and their impact on whitefish (*Coregonus lavaretus* L.). Environmental Toxicology. 2001;**16**:483-488. DOI: 10.1002/tox.10006

[134] Li X-Y, Chung I-K, Kim J-I, Lee J-A. Subchronic oral toxicity of microcystin in common carp (*Cyprinus carpio* L.) exposed to Microcystis under laboratory conditions. Toxicon. 2004;**44**:821-827. DOI: 10.1016/j.toxicon.2004.06.010

[135] Esterhuizen-Londt M, von Schnehen M, Kühn S, Pflugmacher S. Oxidative stress responses in the animal model, Daphnia pulex exposed to a natural bloom extract versus artificial cyanotoxin mixtures. Aquatic Toxicology. 2016;**179**:151-157. DOI: 10.1016/j.aquatox.2016.09.003

[136] Dvorakova D, Dvorakova K, Blaha L, Marsalek B, Knotkova Z. Effects of cyanobacterial biomass and purified microcystins on malformations in *Xenopus laevis*: Teratogenesis assay (FETAX). Environmental Toxicology. 2002;**17**:547-555

[137] Burýšková B, Hilscherová K, Babica P, Vršková D, Maršálek B, Bláha L. Toxicity of complex cyanobacterial samples and their fractions in *Xenopus laevis* embryos and the role of microcystins. Aquatic Toxicology. 2006;**80**:346-354. DOI: 10.1016/j.aquatox.2006.10.001

[138] Osswald J, Carvalho AP, Claro J, Vasconcelos V. Effects of cyanobacterial extracts containing anatoxin-a and of pure anatoxin-a on early developmental stages of carp. Ecotoxicology and Environmental Safety. 2009;**72**:473-478. DOI: 10.1016/j.ecoenv.2008.05.011

[139] Cerbin S, Kraak MHS, de Voogt P, Visser PM, Van Donk E. Combined and single effects of pesticide carbaryl and toxic Microcystis aeruginosa on the life history of Daphnia pulicaria. Hydrobiologia. 2010;**643**:129-138. DOI: 10.1007/s10750-010-0130-1

[140] Cazenave J, Bistoni M d l Á, Zwirnmann E, Wunderlin DA, Wiegand C. Attenuating effects of natural organic matter on microcystin toxicity in zebra fish (*Danio rerio*) embryos—Benefits and costs of microcystin detoxication. Environmental Toxicology. 2006;**21**:22-32. DOI: 10.1002/tox.20151

[141] Magalhães VF, Marinho MM, Domingos P, Oliveira AC, Costa SM, Azevedo LO, et al. Microcystins (cyanobacteria hepatotoxins) bioaccumulation in fish and crustaceans from Sepetiba Bay (Brasil, RJ). Toxicon. 2003;**42**:289-295. DOI: 10.1016/S0041-0101(03)00144-2

[142] Smith JL, Haney JF. Foodweb transfer, accumulation, and depuration of microcystins, a cyanobacterial toxin, in pumpkinseed sunfish (*Lepomis gibbosus*). Toxicon. 2006;**48**:580-589. DOI: 10.1016/j.toxicon.2006.07.009

[143] Papadimitriou T, Kagalou I, Stalikas C, Pilidis G, Leonardos ID. Assessment of microcystin distribution and biomagnification in tissues of aquatic food web compartments from a shallow lake and evaluation of potential risks to public health. Ecotoxicology. 2012;**21**:1155-1166. DOI: 10.1007/s10646-012-0870-y

[144] Liu L-P, Su X-M, Chen T-Y, Li K, Zhan J, Egna H, et al. Evidence of rapid transfer and bioaccumulation of Microcystin-LR poses potential risk to freshwater prawn *Macrobrachiumro senbergii* (de Man). Aquaculture Research. 2016;**47**:3088-3097. DOI: 10.1111/are.12759

[145] Prepas EE, Kotak BG, Campbell LM, Evans JC, Hrudey SE, Holmes CF. Accumulation and elimination of cyanobacterial hepatotoxins by the freshwater clam *Anodontagrandis simpsoniana*. Canadian Journal of Fisheries and Aquatic Sciences. 1997;**54**:41-46. DOI: 10.1139/f96-261

[146] Ibelings BW, Bruning K, de Jonge J, Wolfstein K, Pires LMD, Postma J, et al. Distribution of microcystins in a lake foodweb: No evidence for biomagnification. Microbial Ecology. 2005;**49**:487-500. DOI: 10.1007/s00248-004-0014-x

[147] Xie L, Xie P, Guo L, Li L, Miyabara Y, Park H-D. Organ distribution and bioaccumulation of microcystins in freshwater fish at different trophic levels from the eutrophic Lake Chaohu, China. Environmental Toxicology. 2005;**20**:293-300. DOI: 10.1002/tox.20120

[148] Matsunaga H, Harada K-I, Senma M, Ito Y, Yasuda N, Ushida S, et al. Possible cause of unnatural mass death of wild birds in a pond in Nishinomiya, Japan: Sudden appearance of toxic cyanobacteria. Natural Toxins. 1999;**7**:81-84. DOI: 10.1002/(SICI)1522-7189(199903/04)7:2<81::AID-NT44>3.0.CO;2-O

[149] Chen J, Xie P, Li L, Xu J. First identification of the hepatotoxic microcystins in the serum of a chronically exposed human population together with indication of hepatocellular damage. Toxicological Sciences. 2009;**108**:81-89. DOI: 10.1093/toxsci/kfp009

[150] Chen J, Zhang D, Xie P, Wang Q, Ma Z. Simultaneous determination of microcystin contaminations in various vertebrates (fish, turtle, duck and water bird) from a large eutrophic Chinese lake, Lake Taihu, with toxic Microcystis blooms. Science of the Total Environment. 2009;**407**:3317-3322. DOI: 10.1016/j.scitotenv.2009b.02.005

Chapter 3

Application of Diatom-Based Indices in River Quality Assessment: A Case Study of Lower Ogun River (Abeokuta, Southwestern Nigeria) Using Epilithic Diatoms

Adewole Michael Gbadebo,
Benjamin Onozeyi Dimowo,
Adewale Matthew Taiwo and Isaac Tunde Omoniyi

Additional information is available at the end of the chapter

http://dx.doi.org/10.5772/intechopen.86347

Abstract

Diatom indices have been extensively applied in the bioassessment of surface waters and wetlands in many countries except Nigeria. This pioneer study aimed at investigating the use of epilithic diatom-based indices in the assessment of the Ogun River quality. Water and epilithic diatom samples were collected fortnightly from four sampling stations for a period of four consecutive months (March–June 2015). Water samples were analysed for pH, temperature, electrical conductivity, total dissolved solids (TDS), dissolved oxygen, chemical oxygen demand (COD), nitrite, nitrate, ammonium, phosphate, sulphide, chloride, iron, manganese, silicate, total alkalinity, total hardness, total suspended solids (TSS), transparency and total organic carbon using standard methods. Epilithic diatom samples were collected by scraping the surfaces of rocks or stones using an 18 mm toothbrush and analysed following the standard methods. Data collected were subjected to descriptive (frequency, mean) and inferential statistics (diatom indices, Pearson correlation) using OMNIDIA and SPSS statistical packages. Results showed that the water quality of the Lower Ogun River ranged between bad and high qualities during the study period. The diatom indices (trophic diatom index (TDI), biological diatom index (IBD), generic salinity index (GSI1), generic trophic index (GTI), generic saprobity index (GTI)) were correlated with physical and chemical parameters, thereby indicating their effectiveness in water quality ranking.

Keywords: environmental management, applied ecology, ecosystem health, water quality assessment, surface water bioindicators

1. Introduction

Diatom-based indices are increasingly becoming important tools for the assessment of ecological conditions in lotic and lentic systems [1], and the ability to use diatoms to evaluate present and past conditions of water quality and environmental change in just about any aquatic environment has been recognized worldwide for many decades [2–6].

They are well favoured than other aquatic bioindicators in use during water quality assessment due to their cosmopolitan distribution and well-known ecological requirements. These features enable diatom indices developed in a geographic region to be used in other parts of the world [7]. This explains their wide usage in various water bodies in the world.

Some of the diatom indices in use presently include Descy's index or DES [8], Sládecek's index or SLA [9], Leclercq and Maquet's index or LMI [10], the Watanabe index or WAT [11], the Commission of Economical Community Index or CEC [12], Schiefele and Schreiner's index or SHE [13], Rott's index or ROT [14], the generic diatom index or GDI [15], the Specific Pollution Sensitivity Index or SPI [16], the biological diatom index or IBD [17], the eutrophication/pollution index or EPI [18], the Artois-Picardie Diatom Index or APDI [19], the trophic diatom index or TDI [20], the Pampean Diatom Index or IDP [21] and the South African Diatom Index [22].

These indices according to Taylor et al. [23] function in the following manner: in a sample from a body of water with a particular level of determinant (e.g. salinity), diatom taxa with their optimum close to that level will be the most abundant. Therefore an estimate of the level of that determinant in the sample can be made from the average of the optima of all the taxa in that sample, each weighted by its abundance. A further refinement is the provision of an indicator value which is included to give greater weight to those taxa which are good indicators of particular environmental conditions. In practice, the first step to be completed when using diatom indices is the compilation of a list of taxa in a sample, together with their absolute abundance. The final index value is expressed as the mean of the optima of the taxa in the sample, weighted by the abundance of each taxon. The indicator value acts to further increase the influence of certain species [23, 24].

All these indices are based on the formula of Zelinka and Marvan [25] except the CEC, SHE, TDI and WAT indices [23]. The quality of running rivers has been successfully classified using epilithic diatom-based indices in Bulgaria [26].

Diatom indices have also been utilized in monitoring biotic integrity and trophic condition of aquatic ecosystems in select countries in sub-Saharan Africa [27] which include South Africa [28–35], Kenya [36–39], Zimbabwe [40, 41] and Zambia [42].

However, such studies have not been carried out on Nigerian water bodies. The Ogun River is a perennial water source which passes through areas of high population density [43–45]. In populated areas, river water quality is determined by human practices: deforestation, farming, industrial and domestic sewage discharge, which cause changes in colour, suspended solids, pH, temperature, nutrients and run-off characteristics [46].

http://dx.doi.org/10.5772/intechopen.86347

Being that diatom-based biomonitoring programmes have been implemented with some success in South Africa, Kenya, Zimbabwe and Zambia [27], this study therefore investigated the use of epilithic diatom-based indices in water quality assessment of the Lower Ogun River.

2. Materials and methods

2.1. Description of the study area

Abeokuta is the capital and the largest city in Ogun State which is situated in the southwestern part of Nigeria [47]. Soils in Abeokuta have been characterized as being sandy, formed from sedimentary rocks, and can only support savannah vegetation. Vegetation is predominated by guinea and derived savannah [48]. The Ogun River (**Figure 1**) is one of the main rivers in the southwestern part of Nigeria with a total area of 22.4 km^2 and a fairly large flow of about 393 m^3 s^{-1} during the wet season [49]. It has coordinates of 3028″E and 8041″N from its source in Oyo State to 3025″E and 6035″N in Lagos where it enters the Lagos lagoon [43]. Mean annual rainfall ranges from 900 in the north to 2000 mm towards the south. The estimates of total annual potential evapotranspiration have been put between 1600 and 1900 mm [50]. The Ogun River water is used for agriculture, transportation, human consumption, various industrial activities and domestic purposes [43, 49]. It also serves as a

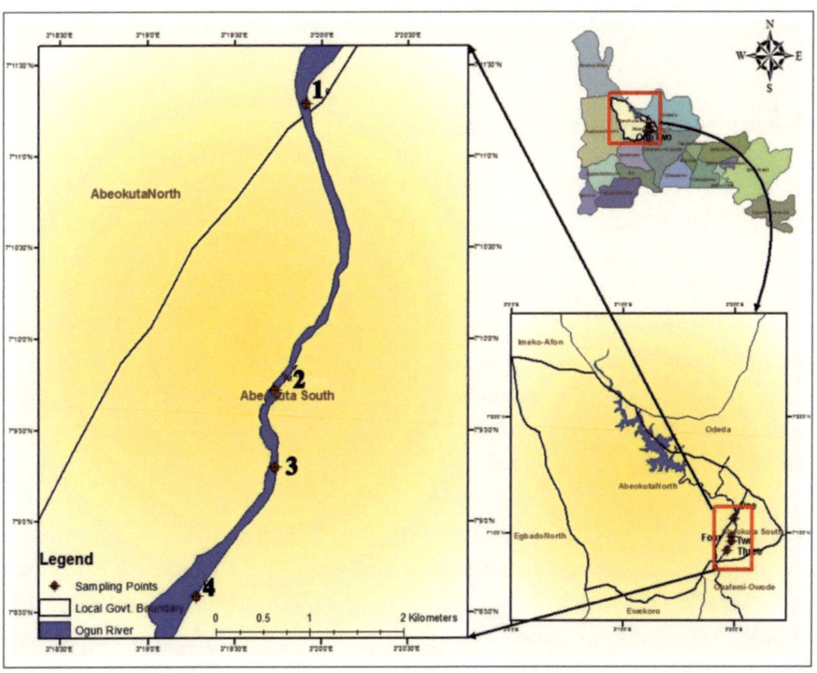

STATIONS	
1	Ibẹrẹkodo
2	Agọ ika
3	Ẹnu gada
4	Off Pepsi Bus Stop

Figure 1. Map of Lower Ogun River, Abeokuta showing the sampling station.

raw material to the Ogun State Water Corporation which treats it before dispensing it to the public. Along its course, it constantly receives effluents from breweries, slaughterhouses, dyeing industries, tanneries and domestic wastewater before finally discharging to Lagos lagoon [43, 45, 49].

Reports on the water quality of the Ogun River have been documented for over 30 years [51, 52]. Several studies have been reported on the water quality of the lower part of the Ogun River at Abeokuta. Among such studies (**Table 1**) include Ojekunle et al. [53], Adeosun et al. [54], Taiwo et al. [55], Olayinka et al. [56], Ikotun et al. [57], Awoyemi [58], Dimowo [44, 45], Osunkiyesi [59] and Adeogun et al. [60].

2.2. Water sampling and analysis procedure

Water samples were collected into well-labelled sample bottles fortnightly for the period of 4 months (March–June 2015) from four sampling stations along the river. Station A was located close to the Ogun State Water Corporation, Arakanga, Ibẹrẹkodo; Station B was located close to the FADAMA III supported ferry at Agọ Ika; Station C was located just below the bridge connecting to Lafenwa at Ẹnu Gada; and Station D was located just down the road of Pepsi bus stop, Quarry Road. The physical and chemical parameters determined included pH, water temperature, electrical conductivity and total dissolved solids which were determined in situ with the use of HANNA Hi 98129 multimeter, while dissolved oxygen, chemical oxygen demand, nitrite, nitrate, ammonium, phosphate, sulphide, chloride, iron, manganese, silicate, total alkalinity, hardness, total suspended solids and total organic carbon were determined in the laboratory using standard methods [61]. Water transparency was also measured in situ using a Secchi disc.

2.3. Epilithic diatom sampling and analysis procedure

Diatom samples were collected fortnightly from four sampling stations along the river for the period of 4 months (March–June 2015). The surfaces of the rocks or stones were scraped off using an 18 mm toothbrush. The brushed areas of the stones, as well as the toothbrush covered with algae, were flushed into a plastic bowl with water. The obtained brown or greenish suspension containing the diatoms was collected and preserved with neutral formaldehyde (4%) to prevent the silica cell walls from cracking. Due to the unavailability of rocks/stones at sampling stations at certain visits, other submerged substrates such as wood material were sampled. This method was adapted from the recommendations of Martin and Fernandez [62] and Kelly et al. [63]. Thereafter, in the laboratory, the samples were mounted on microscope slides by first shaking the samples vigorously and then pipetting a drop onto the slides with the use of a dropper. The identification of the diatoms was done to the lowest taxonomic category possible under the microscope using keys of identification such as [64–67]. Then enumeration was carried out using the drop count method adapted from Dhargalkar and Ingole [68]. The abundance of organisms in each sample was extrapolated from the number of organisms per drop to the number of organisms per ml by multiplying the number of organisms per drop by 20 based on the tested premise that 20 drops of the sample make 1 ml.

http://dx.doi.org/10.5772/intechopen.86347

Parameters	Present study	Ojekunle et al. [53]	Adeosun et al. [54]	Taiwo et al. [55]	Olayinka et al. [56]	Ikotun et al. [57]	Awoyemi [58]	Dimowo [44]	Osunkiyesi [59]	Adeogun et al. [60]
W-T (°C)	25.85–32.3	24.3–27.5	24–30.7	26.8–27	27.87–29.5	23.7–31.7	29–31	26.9–32.1	27–32	24.5–32
TRANS (cm)	14–88.75	NA	53–100	NA	NA	NA	98–173	20–70	NA	NA
COND (μScm^{-1})	127–377	NA	140–190	103.7–105	412.67–514.67	NA	150–388	99–180.5	NA	725.19–3400
TDS (mgL^{-1})	63.5–189	690–7000	70–95	46–48	93.33–95	NA	75–194	48.8–90.8	438–448	346.05–757.03
NO_3^- (mgL^{-1})	0.02–47.5	35–205	0.235–5.445	0.4–0.9	1.85–2.13	0.66–3.91	20.49–63.42	0.6–113.4	NA	12.28–89.43
NO_2^- (mgL^{-1})	0.03–0.4	NA	NA	NA	NA	NA	NA	NA	0.65–0.69	NA
PO_4^- (mgL^{-1})	1.75–28	52–250	0.02–0.75	NA	2.68–3.26	0.19–2.0	0.035–0.583	0–0.1	NA	1.85–18.62
DO (mgL^{-1})	5.51–6.72	0.1–8.82	4.12–5.32	5.5–6.0	3.7–4.75	3.9–7.7	1.88–5.52	2.8–7.7	NA	0–11.27
COD (mgL^{-1})	4–678.5	350–2500	NA	NA	88.33–111.67	NA	NA	NA	NA	181.5–1374.91
pH	8.36–9.91	6.14–7.3	7.45–9.73	7.92–7.96	6.37–7.1	6.5–7.7	6.5–7.95	7.7–9.1	7.6–7.72	5.5–8.8
TSS (mgL^{-1})	0–49	NA	NA	79–95	2.79–5.32	52.9–107.5	NA	NA	446.00–448.09	822.93–1495.47
TA (mgL^{-1})	3.75–10	NA	4.5–14.5	0.1–0.1	NA	NA	NA	4.4–17.8	42.9–43.6	NA
TH (mgL^{-1})	2.6–7.15	NA	NA	41–50	NA	NA	NA	45.5–105	36.1–38.1	NA
Cl^- (mgL^{-1})	25–25	380–1990	NA	NA	NA	29.3–104.5	NA	NA	8.98–9.86	15.33–183.58
Fe^+ (mgL^{-1})	0.05–1.53	NA	NA	0.3–0.4	1.37–1.73	0.12–2.3	NA	NA	13,800–16,100	NA
Mn^+ (mgL^{-1})	0.03–0.66	NA	NA	NA	NA	0–1.0	NA	NA	289–466	NA

W-T = water temperature; pH = hydrogen ion concentration; COND = electrical conductivity; TDS = total dissolved solids; TRANS = water transparency; Fe^+ = iron; NO_2^- = nitrite; NO_3^- = nitrate; Mn^+ = manganese; SiO_3^- = silicate; PO_4 = phosphates; Cl^- = chloride; TA = total alkalinity; TH = total hardness; COD = chemical oxygen demand; TSS = total suspended solids; DO = dissolved oxygen

Table 1. Reports on the water quality of the Lower Ogun river at Abeokuta.

2.4. Statistical analysis

Descriptive statistics in the form of frequency tables and range were used in the presentation of the data. Inferential statistics such as diatom indices, viz. biological diatom index (IBD), trophic diatom index, (TDI) and generic diatom indices (GDI) (saprobity index, trophic index and salinity index) were utilized in determining the water quality status of the Ogun River. IBD was calculated using OMNIDIA free version software [69], while TDI and GDI were calculated using MS Excel spreadsheets [70] following the method adapted from Kelly et al. [20] and Van Dam [71], respectively.

2.4.1. Correlation analysis

Bivariate correlation analysis was carried out using Pearson's product moment coefficient of correlation in SPSS to check for the relationship between the physical and chemical parameters and diatom indices. The physical and chemical parameters (except pH) and diatom abundance data were log transformed before analysis in order to achieve normal distribution.

2.4.2. Ranking of water quality using diatom-based indicators

According to Taylor et al. [23], diatom-based indicators in all cases are calculated using the formula of Zelinka and Marvan [25] except for the Commission of Economical Community Index (CEC), Schiefele and Schreiner's index (SHE), trophic diatom index (TDI) and Watanabe index (WAT index). They have the basic form [4] given:

$$\text{index} = \frac{\sum_j^a = 1^{a_j s_j v_j}}{\sum_j^a = 1^{a_j v_j}} \qquad (1)$$

where a_j is the abundance (proportion) of species j in sample; s_j is the pollution sensitivity of species j; v_j is the indicator value.

The performance of the indices depends on the values given to the constants s and v for each taxon, and the values of the index range from 1 to an upper limit equal to the highest value of s.

Index score	Water quality rank	Trophic status
>17	High quality	Oligotrophy
15–17	Good quality	Oligo-mesotrophy
12–15	Moderate quality	Mesotrophy
9–12	Poor quality	Mesoeutrophy
<9	Bad quality	Eutrophy

Source: Adapted from [73]

Table 2. Water quality ranking with the use of IBD.

Diatom indices differ in the number of species used and in the values of s and v which have been attributed after compiling the data from literature and from ordinations [4, 72]. For all of the above indices, except TDI (maximum value of 100), the maximum value of 5 (converted to 20 by the software package OMNIDIA) indicates a high quality or pristine water resource [23].

The diatom biotic indices, viz. IBD, TDI and GDI, were interpreted following the classifications in **Tables 2–4** as adapted from Kelly et al. [20] and Van Dam [71], Eloranta and Soininen [73] and Delta Environmental [74].

		Percentage of motile valves				**Interpretation**	**Remarks**
		<20%	21 – 40%	41 – 60%	>60%		
TDI Range	0 – 9 10 – 19					A vertical movement on the chart indicates a change in water quality due to nutrients while a horizontal movement indicates change due to other factors.	Oligotrophic Fairly oligotrophic
	20 – 29 30 – 39						Oligo-mesotrophic Highly oligo-mesotrophic
	40 – 49 50 – 59						Fairly mesotrophic Mesotrophic
	60 – 69 70 – 79						Meso-eutrophic Fairly eutrophic
	80 – 89 ≥90						Eutrophic Highly eutrophic

Source: Adapted from [20]

Table 3. TDI water quality lookup chart.

Generic salinity index	Generic trophic index	Generic saprobity index	Water quality classes	Index score
Very clean	Oligotrophic	Oligosaprobous	I	>1
Clean	Oligo-/mesotrophic	β-Mesosaprobous	II	1–0.96
Moderate	Mesotrophic	α-Mesosaprobous	III	0.95–0.76
Polluted	Eutrophic	Meso-/polysaprobous	IV	0.75–0.56
Very polluted	Hypereutrophic	Polysaprobous	V	<0.56

Source: Based on [71, 74]

Table 4. Interpretation of generic diatom indices.

3. Results

The weekly and monthly spatial variation in the physical and chemical parameters and the weekly variation in the epilithic diatom abundance and epilithic diatom indices of the Ogun River at Abeokuta are available as supplementary files.

3.1. Epilithic diatom composition and variation

A total of 61 epilithic diatoms (**Table 5**) belonging to 12 orders and 3 classes were identified in the study sites. *Caloneis bacillum* (3600 cells mL^{-1}) had the highest total count followed by *Coscinodiscus rothii* (2300 cells ml^{-1}) and *Campylodiscus clypeus* (1780 cells mL^{-1}).

The monthly spatial variation in epilithic diatom indices (**Table 6**) was in the following order: Trophic diatom index was found highest in March (69.26; Station C) and lowest in June (14.52; Station C). %Motile taxa was found highest in June (31.42; Station D) and lowest in April (1.10; Station B). Biological diatom index was found highest in June (15.10; Station D) and lowest in June (7.75; Station A). Generic salinity index (GSI1) was found highest in May (20.00; Station B). It was lowest in March, April (0.00; Station C) and June (0.00; Stations B, C, D). Generic trophic index was found highest in June (9.00; Station B) and lowest in June (0.00; Stations C and D). Generic saprobity index was found highest in March (4.00; Station A) and lowest in April (0.08; Station D).

3.2. Water quality parameters

The physical and chemical parameters (**Table 1**) assessed in this study had the following ranges: Water temperature was found highest in May (32.3°C; Station A) and lowest in June (25.85°C; Station D). Water transparency was highest in April (88.75 cm; Station B) and lowest in June (14 cm; Station D). Electrical conductivity was highest in March (377 μS cm^{-1}; Station C) and lowest in March (127 μS cm^{-1}; Station B). Total dissolved solids was highest in March (189 mg L^{-1}; Station C) and lowest in March (63.5 mg L^{-1}; Station B). Nitrate was highest in April (47.5 mg L^{-1}; Station A) and lowest in May (0.02 mg L^{-1}; Station A). Nitrite was highest in March (0.4 mg L^{-1}; Station D) and lowest in June (0.03 mg L^{-1}; Stations A, B and C). Phosphate was highest in April (28 mg L^{-1}; Station C) and lowest in June (1.75 mg L^{-1}; Station D). Dissolved oxygen was highest in June (6.72 mg L^{-1}; Station D) and lowest in March (5.51 mg L^{-1}; Station A). Chemical oxygen demand was found highest in June (678.5 mg L^{-1}; Station D) and lowest in June (4 mg L^{-1}; Station C). pH was highest in April (9.91; Station A) and lowest in March (8.36; Station B). Total suspended solids was highest in May (49 mg L^{-1}; Station D) and lowest in May (0 mg L^{-1}; Station A). Total alkalinity was highest in June (10 mg L^{-1}; Station D) and lowest in March (3.75 mg L^{-1}; Station C). Total hardness was highest in March (7.15 mg L^{-1}; Station D) and lowest in March (2.6 mg L^{-1}; Station A). Chloride levels were constant throughout the study period (25 mg L^{-1}). Iron was found highest in April (1.53 mg L^{-1}; Station C) and lowest in June (0.05 mg L^{-1}; Station A). Manganese was highest in May (0.66 mg L^{-1}; Station D) and lowest in May (0.03 mg L^{-1}; Station B).

Epilithic diatom species	March 2015				April 2015				May 2015				June 2015				Total count
	A	B	C	D	A	B	C	D	A	B	C	D	A	B	C	D	
Melosira varians	—	680	—	—	—	80	20	640	—	120	80	120	—	—	—	20	1760
Aulacoseira granulata	—	40	—	—	300	60	20	60	100	20	220	200	160	100	20	—	1300
Cyclotella meneghiniana	—	—	40	—	60	60	40	—	340	—	—	80	100	20	100	20	860
Cyclotella stelligera	—	240	—	—	—	—	—	—	40	—	—	—	—	—	—	240	520
Coscinodiscus rothii	—	—	80	—	240	120	100	700	220	60	—	100	160	—	340	180	2300
Stephanodiscus margarae	—	—	—	—	—	—	—	—	60	60	—	20	60	—	100	—	300
Stephanodiscus agassizensis	—	—	60	—	—	1160	40	—	180	100	60	100	—	—	—	—	1700
Cyclostephanos tholiformis	—	—	—	—	—	—	—	—	—	40	—	20	—	—	—	—	60
Fragilaria capucina	40	—	—	220	—	—	20	—	—	—	—	20	—	—	—	—	300
Synedra acus	—	—	—	—	—	20	—	—	—	—	—	20	140	—	60	40	280
Synedra nana	—	—	—	—	—	—	—	—	—	—	—	—	20	—	—	20	40
Synedra ulna	60	1140	40	80	160	40	—	80	—	140	220	40	80	40	40	120	2280
Diatoma vulgaris	—	—	100	40	—	—	—	—	—	—	—	—	—	20	—	—	160
Diatoma hiemale	40	—	—	40	—	—	—	—	—	100	20	40	—	—	—	—	240
Diatoma tenuis	—	80	60	40	140	80	100	100	60	20	60	—	—	—	—	—	740
Staurosira construens	—	—	—	—	—	—	—	—	—	—	—	—	60	—	—	—	60
Meridion circulare	—	—	—	—	40	—	—	—	—	40	20	—	—	—	—	—	100
Tetracyclus lacustris	—	20	—	—	—	—	—	—	—	—	—	40	—	—	220	60	340
Cymbella tumida	—	—	—	—	—	—	20	—	20	—	40	20	—	—	—	—	100
Gomphonema parvulum	—	—	—	—	—	—	—	—	—	—	—	—	—	20	—	—	20
Gomphonema acuminatum	—	—	—	—	—	—	—	—	—	—	—	—	20	—	—	—	20
Gomphonema truncatum	—	—	—	—	—	—	—	20	20	—	20	—	—	—	—	—	60

Epilithic diatom species	March 2015				April 2015				May 2015				June 2015				Total count
	A	B	C	D	A	B	C	D	A	B	C	D	A	B	C	D	
Asterionella formosa	60	—	—	—	—	—	—	—	40	—	—	—	—	—	20	—	120
Amphicampa erura	—	—	—	—	—	—	—	—	40	—	—	20	—	—	—	—	60
Eunotia serpentine	60	—	—	—	20	—	—	—	—	40	160	—	—	60	20	—	360
Eunotia formica	20	80	—	—	—	—	—	—	—	—	—	—	—	—	—	—	100
Eunotia bilunaris	—	—	—	—	40	—	—	—	20	—	—	—	—	—	—	—	60
Gyrosigma cf. *scalproides*	20	—	—	—	—	—	—	—	—	—	—	—	—	—	20	—	40
Gyrosigma attenuatum	140	240	—	20	60	—	—	—	20	—	—	—	40	40	—	—	560
Navicula viridula	—	—	—	—	—	—	—	—	—	—	—	—	—	40	60	—	100
Navicula radiosa	20	500	—	—	—	—	—	—	—	—	—	—	—	—	—	20	540
Navicula capitoradiata	140	—	200	—	—	—	—	—	—	—	—	—	—	—	—	—	340
Navicula cryptocephala	20	140	120	—	—	—	20	—	—	—	—	—	—	—	—	—	300
Navicula spp.	—	—	—	—	60	—	—	—	—	—	—	—	—	—	—	—	60
Navicula cf. *margalithi*	—	280	380	—	—	—	—	—	—	—	—	—	—	—	—	—	660
Pinnularia viridis	40	140	—	40	260	—	—	—	180	320	—	—	—	—	—	—	980
Pinnularia cf. *interrupta*	20	40	—	—	60	—	40	—	40	20	40	20	—	—	—	—	280
Frustulia vulgaris	—	—	—	—	—	20	—	—	40	—	40	120	20	60	80	420	800
Sellaphora pupula	—	—	20	—	—	—	—	—	—	—	—	—	—	—	—	—	20
Sellaphora seminulum	—	—	—	—	—	40	—	20	—	—	—	—	—	—	—	—	60
Sellaphora bacillum	—	—	—	20	20	60	—	—	—	—	—	—	—	—	—	—	100
Stauroneis cf. *kriegerii*	—	—	20	—	—	—	40	—	—	—	—	—	—	—	—	—	60
Caloneis bacillum	—	—	—	—	—	—	—	—	—	—	—	—	—	—	—	3600	3600
Craticula cuspidata	—	140	—	—	—	—	—	—	—	—	—	—	—	—	—	—	140

http://dx.doi.org/10.5772/intechopen.86347

Epilithic diatom species	March 2015				April 2015				May 2015				June 2015				Total count
	A	B	C	D	A	B	C	D	A	B	C	D	A	B	C	D	
Achnanthes lanceolata	40	—	20	40	—	—	—	—	—	—	—	80	40	—	—	—	220
Planothidium lanceolatum	—	—	—	—	—	—	—	—	—	—	—	20	—	—	—	—	20
Achnanthes cf. *inflate*	—	—	—	—	—	—	—	—	—	—	—	—	—	20	—	—	20
Cocconeis placentula	—	—	—	—	—	—	—	—	—	—	40	—	40	—	—	40	120
Nitzschia cf. *acicularis*	—	—	—	—	—	—	—	—	—	—	—	20	—	—	—	—	20
Nitzschia cf. *dissipata*	20	—	—	—	—	—	—	—	—	—	—	—	20	—	—	—	40
Nitzschia palea	180	180	—	60	40	—	—	100	—	—	—	—	—	—	—	—	560
Nitzschia intermedia	500	—	—	—	—	—	—	—	—	—	—	—	—	—	—	—	500
Bacillaria paradoxa	—	—	—	—	—	—	—	—	40	—	—	—	—	—	—	—	40
Cylindrotheca gracilis	—	—	—	—	—	—	20	—	—	—	—	20	—	—	—	—	40
Cymatopleura solea	—	—	—	—	—	—	20	—	—	—	—	—	—	—	—	—	20
Campylodiscus clypeus	80	—	—	—	120	60	—	20	80	—	200	200	520	80	420	—	1780
Rhopalodia gibba	—	—	20	—	120	20	—	20	—	—	—	—	—	—	—	—	180
Epithemia adnata	—	—	—	—	—	—	—	—	—	—	20	—	20	—	60	—	100
Epithemia sorex	—	—	—	—	—	—	—	—	—	—	—	—	—	—	180	—	180
Denticula subtilis	—	—	—	—	—	—	—	—	40	—	—	—	—	—	60	20	120
Amphora veneta	20	—	—	—	40	20	—	—	40	20	40	—	—	—	—	—	180

Table 5. Monthly spatial variation in the count of epilithic diatoms of the Ogun River in Abeokuta (cells mL^{-1}).

Epilithic diatom indices	March 2015				April 2015				May 2015				June 2015			
	A	**B**	**C**	**D**	**A**	**B**	**C**	**D**	**A**	**B**	**C**	**D**	**A**	**B**	**C**	**D**
TDI	60.44	60.23	69.26	46.30	21.19	56.25	37.96	68.60	15.09	18.61	25.68	47.40	66.91	43.33	14.52	41.15
%Motile taxa	15.63	9.40	12.90	3.90	10.82	1.10	7.35	1.55	15.05	10.83	2.67	5.76	3.20	8.54	8.16	31.42
Interpretation	P	P	P	M	G	M	G	P	H	H	G	M	P	M	H	M
IBD	11.47	11.79	9.04	14.90	11.19	10.08	11.12	9.17	10.08	13.36	11.70	11.04	7.75	12.04	7.77	15.10
Interpretation	P	P	P	M	P	P	P	P	P	M	P	P	B	M	B	G
GSI1	0.25	1.67	0.00	1.33	8.50	3.00	0.00	0.80	9.50	20.00	12.00	17.00	11.00	0.00	0.00	0.00
Interpretation	B	H	B	H	H	H	B	M	H	H	H	H	H	B	B	B
GTI	0.24	1.15	0.25	0.44	3.09	0.60	1.20	0.30	3.60	2.86	2.88	5.67	3.33	9.00	0.00	0.00
Interpretation	B	H	B	B	H	P	H	B	H	H	H	H	H	H	B	B
GSI2	3.00	0.30	0.20	0.67	2.83	0.25	0.75	0.08	0.75	1.54	2.50	1.42	2.40	4.50	0.40	0.14
Interpretation	H	B	B	P	H	B	P	B	P	H	H	H	H	H	B	B

TDI = trophic diatom index; %MT = %motile taxa; IBD = biological diatom index; GSI1 = generic salinity index; GTI = generic trophic index; GSI2 = generic saprobity index; H = high quality; G = good quality; M = moderate quality; P = poor quality; B = bad quality

Table 6. Monthly spatial variation in the epilithic diatom indices of Ogun River, Abeokuta.

Parameters	TDI	%Motile taxa	IBD	GSI1	GTI	GSI2
LogWT	0.068	−0.416	−0.120	0.187	−0.298	−0.251
pH	−0.153	−0.295	−0.268	0.383	−0.003	0.161
LogCOND	0.055	0.341	−0.035	−0.301	−0.199	−0.073
LogTDS	0.081	0.372	0.003	−0.295	−0.196	−0.070
LogTRANS	−0.101	−0.495	−0.239	0.322	−0.098	−0.014
LogFe	0.261	−0.168	−0.041	−0.040	−0.148	−0.515*
LogNO_2	−0.205	−0.138	0.512*	0.362	−0.031	−0.146
LogNO_3	0.443	−0.228	−0.090	−0.554*	−0.488	−0.236
LogMn	0.404	−0.342	−0.233	0.025	0.132	−0.107
LogNH_4	0.206	−0.466	−0.046	−0.345	−0.394	−0.513*
LogSO_3	0.397	−0.376	−0.276	−0.206	−0.257	−0.333
LogSiO_3	0.508*	0.058	0.167	−0.470	−0.379	−00.069
LogPO_4	0.109	−0.322	0.033	0.013	−0.222	−0.240
LogCl	.[a]	.[a]	.[a]	.[a]	.[a]	.[a]
LogTA	−0.280	0.079	−0.038	0.098	0.182	0.281
LogTOC	−0.490	0.161	−0.015	0.229	0.363	0.221
LogTH	−0.089	−0.042	0.244	0.121	0.060	−0.502*
LogCOD	0.406	0.185	0.279	0.273	−0.106	−0.217
LogTSS	0.253	−0.099	−0.095	0.361	0.099	0.013
LogDO	−0.009	0.107	0.101	−0.133	−0.210	−0.400
TDI	1	−0.181	−0.187	−0.374	−0.206	−0.130
%Motile taxa	−0.181	1	0.429	−0.189	−0.162	−0.090
IBD	−0.187	0.429	1	0.034	0.033	0.067
GSI1	−0.374	−0.189	0.034	1	0.424	0.225
GTI	−0.206	−0.162	0.033	0.424	1	0.716**
GSI2	−0.130	−0.090	0.067	0.225	0.716**	1

*Correlation is significant at the 0.05 level (two-tailed).

**Correlation is significant at the 0.01 level (two-tailed).

LogWT = log water temperature; pH = hydrogen ion concentration; LogCOND = log electrical conductivity; LogTDS = log total dissolved solids; LogTRANS = log water transparency; LogFe = log iron; LogNO_2 = log nitrite; LogNO_3 = log nitrate; LogMn = log manganese; Log NH_4 = log ammonium; LogSO_3 = log sulphide; LogSiO_3 = log silicate; LogPO_4 = log phosphate; LogTA = log total alkalinity; LogTOC = log total organic carbon; LogTH = log total hardness; LogCOD = log chemical oxygen demand; LogTSS = log total suspended solids; LogDO = log dissolved oxygen; TDI = trophic diatom index; IBD = biological diatom index; GSI1 = generic salinity index; GTI = generic trophic index; GSI2 = generic saprobity index

Table 7. Pearson correlation coefficients of physical and chemical parameters and epilithic diatom indices of the Ogun River at Abeokuta.

3.3. Relationship between physical and chemical parameters and epilithic diatom indices

Table 7 shows the Pearson correlation coefficients of physical and chemical parameters and epilithic diatom indices of the Ogun River at Abeokuta. Physical and chemical parameters and epilithic diatom indices exhibited the following relationship pattern: trophic diatom index was positively correlated with silicate ($r = 0.508$; $p < 0.05$). Biological diatom index was positively correlated with nitrite ($r = 0.512$; $p < 0.05$). Generic salinity index was negatively correlated with nitrate ($r = -0.554$; $p < 0.05$). Generic trophic index was positively correlated with generic saprobity index ($r = 0.716$; $p < 0.01$). Generic saprobity index was negatively correlated with iron ($r = -0.515$; $p < 0.05$), ammonium ($r = -0.513$; $p < 0.05$) and total hardness ($r = -0.502$; $p < 0.05$).

4. Discussion

Diatoms have been shown through research to be used as alternative/supplementary means of water quality assessment due to the specific water quality tolerance each species portend [4, 75–76]. They are sensitive and strongly respond to physicochemical and biological changes [77]. More so, the use of diatoms in water quality assessment is cheaper than routine chemical analyses and directly shows the impact of pollution on the aquatic biota [23].

The range of values got from this study on the physical and chemical parameters was comparable with those reported by previous studies except for pH which was more basic and total hardness which was relatively lower than the range of values previously reported [44, 45, 53–60].

A total of 61 diatom species were identified in this study. *Caloneis bacillum* emerged with the highest total count (cells mL^{-1}) followed by *Coscinodiscus rothii* and *Campylodiscus clypeus*. The dominance of *Caloneis bacillum* has also been reported in Lake Tanganyika by Cocquyt [78]. *Caloneis bacillum* has been reported by Ali et al. [79] to be present in Upper Dilimi River in Jos. *Caloneis bacillum* has also been reported to be dominant by Compere [80] at the fourth sampling site of the Red Sea Hills in north-eastern Sudan.

Following the ecological indicator values reported by Van dam et al. [71], *Caloneis bacillum* occurs mainly in water bodies but is sometimes found in wet environments. It is a nitrogen-autotrophic taxa, tolerating very small amounts of organically bound nitrogen with fairly high oxygen requirements (>75% saturation). It is rarely found in large numbers in rivers [81]. *Caloneis bacillum* is alkaliphilous mainly occurring in fresh brackish waters with $pH > 7$, chloride levels <500 mg L^{-1} and salinity <0.9%. *Caloneis bacillum* is meso-eutraphentic and β-mesosaprobous, thereby falling under water quality class II [71]. It has also been reported to be ubiquitous [82].

The dominance of *Caloneis bacillum* in this study was therefore indicative of moderate pollution.

All the diatom indices (TDI, IBD, GSI1, GSI2, GTI) differed in their ranking of the water quality of the Lower Ogun River at Abeokuta. However, the generic diatom indices (GSI1, GSI2, GTI) were quite similar in their water quality ranking. This misalliance is explained by the global nature of indices, which try to evaluate the general state of water quality and not only the trophic degree [83, 84].

The trophic diatom index (TDI) showed that during the study period, the river water was in most cases poor and moderate (there was a tie in frequency of occurrence) in terms of quality. However, the biological diatom index (IBD) showed that the river water was in most cases moderate in terms of quality.

The generic salinity index (GSI1) showed that the river water was in most cases high in terms of quality during the study period. This salinity classification was calculated based on the tolerance of diatoms to salinity.

The generic trophic index (GTI) ranked the river water during the study period as being in most cases high in terms of quality. This trophic classification was calculated based on the tolerance of diatoms to the trophic state of the aquatic ecosystem. According to Naumann [85] as cited by Van dam et al. [71], variations in trophic state are usually as a result of variations in concentration of inorganic nitrogen and phosphorus compounds. There are however various concepts regarding trophic state. For this reason, water quality assessment based on trophic state was rather qualitative.

The generic saprobity index (GSI2) showed that the river water was in most cases high in terms of quality during the study period. The saprobity classification was calculated based on the indicator properties of diatoms to the presence of biodegradable organic matter and oxygen concentrations in the aquatic ecosystem [71].

Diatom species react distinctly to varying physical and chemical parameters. They are sensitive to change in nutrient concentrations, supply rates and silica/phosphate ratios. Each taxon has a specific optimum and tolerance for nutrients such as phosphate and nitrogen, and this is usually quantifiable [4].

The following deductions were made from the relationship between physical and chemical parameters and epilithic diatom indices: As trophic diatom index (TDI) scores increased, the concentration of silicates increased. This shows that the increases and decreases in concentration of silicates in the river water supported the concomitant increase and decrease in TDI scores. This result corroborated Reynolds [86] as cited by Gbadebo et al. [87] who observed that silica plays an important role in the ecology of aquatic systems as it is an essential element for diatom existence comprising 26–69% of its cellular dry weight. This study did not observe correlations between TDI and pH as observed by Tan et al. [88] in South-East Queensland River, Australia. This study did not also observe correlations between TDI and total phosphates as observed by Vilbaste [89].

It was observed that biological diatom index (IBD) increased as the concentration of nitrites increased. This shows that nitrite influenced the diatoms of the aquatic ecosystem which was evidenced in the IBD scores. Nitrites are one of the nutrients that favour the growth of diatoms.

This result was supported by Kalyoncu and Şerbetci [1] who reported significant correlations between IBD, dissolved oxygen, temperature, conductivity, ammoniacal nitrogen, nitrite nitrogen and phosphate phosphorus. This result was also corroborated by Vilbaste [89] who reported significant correlations of IBD with water temperature, pH, total phosphate, nitrite nitrogen and ammoniacal nitrogen. This study however did not observe correlations between IBD and pH as observed by Tan et al. [88] in Upper Han River, China.

The lack of significant relationship between TDI, IBD, electrical conductivity and total dissolved solids was supported by the observation of Eassa [90] who reported that TDI showed no significant correlation with any physicochemical parameters and/or percentages of eutrophic species. This however did not corroborate Solak et al. [91] who reported negative correlations between TDI, electrical conductivity and total dissolved solids.

The relative abundance of motile diatom taxa in this study did not exhibit significant relationship with the physical and chemical parameters.

The relationship between the generic salinity index (GSI1) scores and the concentration of nitrates in the river water signified that increases in nitrate contributed to reduction in GSI1 scores, whereas no significant relationship was observed between generic trophic index (GTI) scores and physical and chemical parameters except with generic saprobity index (GSI2) scores. Also, generic saprobity index (GSI2) scores decreased as iron, ammonium and total hardness increased but increased with increasing GTI scores.

These results show that there was a close relationship between physical and chemical parameters and diatom-based indices. This is agreed with Bere et al. [40] who applied the indices to urban streams in Zimbabwe. The lack of significant correlation observed between electrical conductivity and the diatom indices in this study was not in agreement with the work of Stancheva et al. [26] who reported high negative correlation.

5. Conclusion

The diatom indices except the relative abundance of motile taxa were moderately correlated with physical and chemical parameters indicating their effectiveness in water quality ranking.

The water quality of the Ogun River during the study period as elucidated from the diatom indices ranged between bad and high qualities. Trophic diatom index (TDI) served as an indicator of silicates, biological diatom index (IBD) was an indicator of nitrites, generic salinity index (GSI1) was an indicator of nitrates and generic saprobity index (GSI2) was an indicator of iron, ammonium and total hardness. It can be concluded that trophic diatom index, biological diatom index, generic salinity index and generic saprobity index can be utilized in water quality assessment.

Limitations on the use of diatom indices in Nigeria include insufficient information on the autecology of diatom species in Nigeria. It is therefore recommended that the ecology of diatoms in Nigeria should be studied in detail in order to provide information on taxonomy, nomenclature, autecology, sensitivities and tolerance levels of diatoms to pollution in

Nigerian waters. Also, diatom keys, identification guides and diatom-based indices specific to water bodies in Nigeria should be developed just as is done in other regions of the world.

Acknowledgements

We acknowledge the Departments of Aquaculture/Fisheries Management and Environmental Management/Toxicology at the Federal University of Agriculture, Abeokuta, Nigeria, for letting us use the research facilities. We are deeply grateful to Prof. (Mrs) M.O. Kadiri of the University of Benin, Nigeria, and Dr. J.C. Taylor of the Northwest University, South Africa, for their support in terms of research materials. We also acknowledge the many unnamed persons that rendered support.

Conflict of interest

The authors declare no conflicts of interest.

Author details

Adewole Michael Gbadebo[1], Benjamin Onozeyi Dimowo[1]*, Adewale Matthew Taiwo[1] and Isaac Tunde Omoniyi[2]

*Address all correspondence to: benjamindimowo@yahoo.com

1 Department of Environmental Management and Toxicology, Federal University of Agriculture, Abeokuta, Nigeria

2 Department of Aquaculture and Fisheries Management, Federal University of Agriculture, Abeokuta, Nigeria

References

[1] Kalyoncu H, Şerbetci B. Applicability of diatom-based water quality assessment indices in Dari stream, Isparta-Turkey. International Journal of Environmental, Ecological, Geological and Mining Engineering. 2013;**7**(6):191-199

[2] Chessman BC. Diatom flora of an Australian river system: Spatial patterns and environmental relationships. Freshwater Biology. 1986;**16**:805-819

[3] Whitmore TJ. Florida diatom assemblages as indicators of trophic state and pH. Limnology and Oceanography. 1989;**34**(5):882-895

[4] Harding WR, Archibald CGM, Taylor JC. The relevance of diatoms for water quality assessment in South Africa: A position paper. Water SA. 2005;**31**(1):41-46

[5] Ács É, Borics G, Fehér G, Kiss KT, Reskóné NM, Stenger-Kovác CS, et al. Implementation of the European water framework directive to assessment the water quality of Hungarian running waters with diatoms. In: Van Dam, Cadee, editors. Proceedings of the Joint Meeting of the Dutch-Flemish Society of Diatomists (NVKD) and 3rd Central European Diatom Meeting (CE-DiatoM) 26-29 March 2009. Diatomededelingen 33. 2009. pp. 29-32

[6] Gottschalk S, Kahlert M. Littoral diatoms as indicators for water quality in Swedish lakes. In: Van Dam, Cadee, editors. Proceedings of the Joint Meeting of the Dutch-Flemish Society of Diatomists (NVKD) and 3rd Central European Diatom Meeting (CE-DiatoM) 26-29 March. 2009. Diatomededelingen 33. 2009. pp. 58-58

[7] Alvarez-Blanco I, Cejudo-Figueiras C, Blanco S, Hernández N, Bécares E. Autoecology of epilithic diatoms in the Duero rivers basin. In: Van Dam, Cadee, editors. Proceedings of the Joint Meeting of the Dutch-Flemish Society of Diatomists (NVKD) and 3rd Central European Diatom Meeting (CE-DiatoM) 26-29 March 2009. Diatomededelingen 33. 2009. pp. 33-33

[8] Descy JP. A new approach to water quality estimation using diatoms. Nova Hedwigia. 1979;**64**:305-323

[9] Slàdecek V. Diatoms as indicators of organic pollution. Acta Hydrochimica et Hydrobiologica. 1986;**14**(5):555-566

[10] Leclerq L, Maquet B. Two new chemical and diatomic indices of running water quality, Application to Samson and its tributaries (Belgian Meuse basin), Comparison with other chemical, biocenotic and diatomic indices, Royal Institute of Natural Sciences of Belgium. Work documents, Belgium. 1987;**38**:1-113

[11] Watanabe T, Asai K, Houki A. Numerical simulation of organic pollution in flowing waters. Encyclopedia of Environmental Control Technology. 1990;**4**:251-281

[12] Descy JP, Coste M. A test of methods for assessing water quality based on diatoms. Verhandlungen des Internationalen Verein Limnologie. 1991;**24**:2112-2116

[13] Schiefele S, Schreiner C. The use of diatoms for monitoring nutrient enrichment, acidification and impact of salt in rivers in Germany and Austria. In: Witton BA, Rott E, Friedrich G, editors. Use of Algae for Monitoring Rivers. Universität Innsbruck: Institut für Botanik; 1991. pp. 103-110

[14] Rott E. Methodological aspects and perspectives in the use of periphyton for monitoring and protecting rivers. In: Witton BA, Rott E, Friedrich G, editors. Use of Algae for Monitoring Rivers. Universität Innsbruck: Institut für Botanik; 1991. pp. 9-16

[15] Coste M, Ayphassorho. Étude de la qualité Des Eaux du Bassin Artois-Picardie à l'aide Des communautés de diatomées Benthiques (Application Des Indices Diatomiques). Rapport CEMAGREF. Bordeaux, Douai: Agence de l'Eau Artois-Picardie; 1991

[16] CEMAGREF. Etude Des méthodes Biologiques Quantitatives d'appréciation de la qualité Des Eaux. Pierre-Bénite: Rapport Division Qualité des Eaux Lyon–Agence Financière de Bassin Rhône-Méditerranée-Corse; 1982

[17] Lenoir A, Coste M. Development of a practical diatom index of overall water quality applicable to the French National Water Board net work. In: Whitton BA, Rott E, editors. Use of Algae for Monitoring Rivers II. Universität Innsbruck: Institut für Botanik; 1996. pp. 29-43

[18] Dell'Uomo A. Assessment of water quality of an Apennine river as a pilot study. In: Whitton BA, Rott E, editors. Use of Algae for Monitoring Rivers II. Universität Innsbruck: Institut für Botanik; 1996. pp. 65-73

[19] Prygiel J, Lévêque L, Iserentant R. Un nouvel in dice diatomique pratique pour l'évaluation de la qualité des eaux en réseau de surveillance. Revue des Sciences de l'Eau. 1996;**1**:97-113

[20] Kelly MG, Adams C, Graves AC, Jamieson J, Krokowski J, Lycett EB, et al. The trophic diatom index: A user's manual, revised edition, R&D technical report, E2/TR2. Environment Agency. 2001. 135pp

[21] Gómez N, Licursi M. The Pampean diatom index (IDP) for assessment of rivers and streams in Argentina. Aquatic Ecology. 2001;**35**:173-181

[22] Harding WR, Taylor JC. The South African Diatom Index (SADI): A preliminary index for indicating water quality in rivers and streams in Southern Africa. In: Report to the Water Research Commission. 2011. 113pp

[23] Taylor JC, Harding WR, Archibald CGM, Rensburg L. Diatoms as indicators of water quality in the Jukskei-Crocodile river system in 1956 and 1957, a re-analysis of diatom count data generated by BJ Cholnoky. Water SA. 2005;**31**(2):237-246

[24] Kelly MG. Use of community-based indices to monitor eutrophication in European rivers. Environmental Conservation. 1998;**25**:22-29

[25] Zelinka M, Marvan P. Zur Präzisierung der biologischen Klassifikation der Reinheit fliessender Gewässer. Archiv für Hydrobiologie. 1961;**57**:389-407

[26] Stancheva R, Ivanov P, Mancheva A. Application of epilithic diatoms in water quality assessment of the rivers vit and Osum, Bulgaria. In: Van Dam, Cadee, editors. Proceedings of the Joint Meeting of the Dutch-Flemish Society of Diatomists (NVKD) and 3rd Central European Diatom Meeting (CE-DiatoM) 26-29 March. 2009. Diatomededelingen 33. 2009. pp. 114-118

[27] Dalu T, Froneman PW. Diatom-based water quality monitoring in southern Africa: Challenges and future prospects. Water SA. 2016;**42**(4):551-559

[28] Schoeman FR. Diatoms as indicators of water quality in the upper Hennops River. Journal of the Limnological Society of Southern Africa. 1979;**5**:73-78

[29] Schoeman FR, Haworth EY. Diatom as indicator of pollution. In: Proceedings of the 8th International Diatom Symposium, Koeltz Scientific Books, Koenigstein. 1986. pp. 757-766

[30] Pieterse AJH, Van Zyl JM. Observations on the relation between phytoplankton diversity and environmental factors in the Vaal River at Balkfontein, South Africa. Hydrobiologia. 1988;**169**:199-207

[31] De La Rey PA, Taylor JC, Laas A, Van Rensburg L, Vosloo A. Determining the possible application value of diatoms as indicators of general water quality: A comparison with SASS 5. Water SA. 2004;**30**:325-332

[32] De La Rey PA, Roux H, Van Rensburg L, Vosloo A. On the use of diatom-based biological monitoring. Part 2: A comparison of the response of SASS 5 and diatom indices to water quality and habitat variation. Water SA. 2008;**34**:61-70

[33] Harding WR, Archibald CGM, Taylor JC, Mundree S. The South African diatom collection: An appraisal and overview of needs and opportunities. In: WRC Report No. TT 242/04. Pretoria: Water Research Commission; 2004

[34] Taylor JC, Janse Van Vuuren MS, Pieterse AJH. The application and testing of diatom-based indices in the Vaal and Wilge rivers, South Africa. Water SA. 2007;**33**:51-60

[35] Taylor JC, Prygiel J, Vosloo A, De La Rey PA, Van Rensburg L. Can diatom-based pollution indices be used for biomonitoring in South Africa? A case study of the Crocodile West and Marico water management area. Hydrobiologia. 2007;**592**:455-464

[36] Ndiritu GG, Gichuki NN, Kaur P, Triest L. Characterization of environmental gradients using physico-chemical measurements and diatom densities in Nairobi River, Kenya. Aquatic Ecosystem Health & Management. 2003;**6**:343-354

[37] Ndaruga AM, Ndiritu GG, Gichuki NN, Wamicha WN. Impact of water quality on macroinvertebrates assemblages along a tropical stream in Kenya. African Journal of Ecology. 2004;**42**:208-216

[38] Ndiritu GG, Gichuki NN, Triest L. Distribution of epilithic diatoms in response to environmental conditions in an urban tropical stream, Central Kenya. Biodiversity and Conservation. 2006;**15**:3267-3293

[39] Triest L, Lung'ayia H, Ndiritu G, Beyene A. Epilithic diatoms as indicators in tropical African rivers (Lake Victoria catchment). Hydrobiologia. 2012;**695**:343-360

[40] Bere T, Mangadze T, Mwedzi T. The application and testing of diatom-based indices of stream water quality in Chinhoyi town, Zimbabwe. Water SA. 2014;**40**(3):503-512

[41] Bere T, Mangadze T. Diatom communities in streams draining urban areas: Community structure in relation to environmental variables. Tropical Ecology. 2014;**55**:271-281

[42] Lang P, Taylor JC, Bertolli L, Lowe S, Dallas H, Kennedy MP, et al. Proposed procedure for the sampling, preparation and analysis of benthic diatoms from Zambian rivers: A bioassessment and decision support tool applicable to freshwater ecoregions in tropical southern Africa. In: Southern African River assessment scheme (SAFRASS). AFS/2009/219013. ACP Science and Technology Programme, Cape Town. 2012

[43] Ayoade AA, Sowunmi AA, Nwachukwu HI. Gill asymmetry in *Labeo ogunensis* from Ogun River, Southwest Nigeria. Revista de Biología Tropical. 2004;**52**(1):171-175

http://dx.doi.org/10.5772/intechopen.86347

[44] Dimowo BO. An assessment of the physico-chemical parameters and plankton composition of Ogun River, Southwest Nigeria. In: BSc. Report. Abeokuta, Ogun State, Nigeria: Federal University of Agriculture; 2012. 97 pp

[45] Dimowo BO. Monthly spatial occurrence of phytoplankton and zooplankton in River Ogun, Abeokuta, Ogun State, Southwest Nigeria. International Journal of Fisheries and Aquaculture. 2013;**5**(8):193-203

[46] Silva-Benavides A. The use of water chemistry and benthic diatom communities for qualification of a polluted tropical river in Costa Rica. Revista De Biologia Tropical. 1995;**44**(2):395-416

[47] National Bureau of Statistics (NBS). State Information (Online). www.nigerianstat.gov.ng/information/details/Ogun (25 Jan. 2015); 2012

[48] Online Nigeria, 2016. Physical setting of Ogun state (online). Available from: http://links.onlinenigeria.com/ogunadv.asp?Blurb=337 [20 March 2016]

[49] Oketola AA, Osibanjo O, Ejelonu BC, Oladimeji YB, Damazio OA. Water quality assessment of river Ogun around the cattle market of Isheri. Nigerian Journal of Applied Science. 2006;**6**:511-517

[50] Bhattacharya AK, Bolaji GA. Fluid flow interactions in Ogun River, Nigeria. International Journal of Research and Reviews in Applied Sciences. 2010;**2**(2):173-180

[51] Adebisi AA. The physico-chemical hydrology of tropical seasonal river—Upper Ogun River. Hydrobiology. 1981;**79**:157-165

[52] Martins O. The Ogun River: Geochemical characteristics of a small drainage basin. Hydrological Processes. 1987;**64**:475-482

[53] Ojekunle ZO, Ufoegbune GC, Oyebamiji FF, Sangowusi RO, Taiwo AM, Ojekunle VO. Assessment of the effect of commercial activities on the surface water quality of Ogun River, Nigeria. Merit Research Journal of Environmental Science and Toxicology. 2014;**2**:196-204

[54] Adeosun FI, Akin-Obasola BJ, Oyekanmi FB, Kayode JO. Physical and chemical parameters of lower Ogun River Akomoje, Ogun State, Nigeria. Fisheries and Aquaculture Journal. 2014;**5**:88-98

[55] Taiwo AG, Adewunmi AR, Oseni OA, Ajayi JO, Lanre-Iyanda YA. Physico-chemical and microbial analysis of the impact of abattoir effluents on Ogun River course. International Journal of ChemTech Research. 2014;**6**(5):3083-3090

[56] Olayinka OO, Adedeji OH, Oladeru IB. Water quality and bacteriological assessment of slaughterhouse effluent on urban river in Nigeria. Journal of Applied Sciences in Environmental Sanitation. 2013;**8**(4):277-286

[57] Ikotun OO, Olafusi OS, Quadri HA, Bolarinwa OA. Influence of human activities on the water quality of Ogun River in Nigeria. Civil and Environmental Research. 2012;**2**:36-48

[58] Awoyemi AO. Physical and chemical parameters of Ogun river (Opeji river), Opeji Village, Abeokuta, Ogun State, Nigeria. In: BSc. Report. Abeokuta, Ogun State, Nigeria: Federal University of Agriculture; 2012

[59] Osunkiyesi AA. Physico-chemical analysis of Ogun River (water samples) within two locations (Akin Olugbade and Lafenwa) in Abeokuta, Ogun State, Nigeria. IOSR Journal of Applied Chemistry. 2012;**1**(4):24-27

[60] Adeogun AO, Chukwuka AV, Ibor OR. Impact of Abattoir wastes and saw-mill effluents on water quality of Upper Ogun River (Abeokuta). American Journal of Environmental Services. 2011;**7**(6):525-530

[61] Merck. Spectroquant Nova 60 Operating manual. Darmstadt, Germany: Merck KGaA; 2014. 297 pp

[62] Martin G, Fernandez M. Diatoms as indicators of water quality and ecological status: Sampling, analysis and some ecological remarks. In: Voudouris, editor. Ecological Water Quality–Water Treatment and Reuse. United Kingdom: InTech; 2012. pp. 183-204

[63] Kelly MG, Penny CJ, Whitton BA. Comparative performance of benthic diatom indices used to assess river water quality. Hydrobiologia. 1995;**302**:179-188

[64] Edmondson WT, editor. Freshwater Biology. New York: John Wiley Sons Inc; 1959. ISBN: 0-471-23298-X

[65] Gell PA, Sonneman JA, Reid MA, Illman MA, Sincock AJ. An illustrated key to common diatom genera from Southern Australia. In: Identification Guide No. 26. Cooperative Research Centre for Freshwater Ecology; 1999. 63 pp

[66] Biggs BJF, Kilroy C. Stream Periphyton Monitoring Manual. New Zealand: NIWA–New Zealand Ministry for the Environment; 2000. 246 pp

[67] Janse van vuuren S, Taylor J, Van Ginkel C, Gerber A. Easy identification of the most common freshwater algae: A guide for the identification of microscopic algae in South African freshwaters. IBM Corporation. 2006;**2011**

[68] Dhargalkar VK, Ingole BS. Measurement of Biomass; Phytoplankton Identification Manual. NIWA–New Zealand Ministry for the Environment; 2004. pp. 25-26

[69] Lecointe C, Coste M, Prygiel J. OMNIDIA: Software for taxonomy, calculation of diatom indices and inventories management. Hydrobiologia. 1993;**269**(270):509-513

[70] Microsoft Corporation. Microsoft Excel, Version 2007, Microsoft Office Suite, DVD; 2006

[71] Van Dam H, Mertens A, Sinkeldam J. A coded checklist and ecological indicator values of freshwater diatoms from the Netherlands. Netherlands Journal of Aquatic Ecology. 1994;**28**(1):117-133

[72] Prygiel J, Coste M. The assessment of water quality in the Artois-Picardie water basin (France) by the use of diatom indices. Hydrobiologia. 1993;**269**(279):343-349

[73] Eloranta P, Soininen J. Ecological studies of some Finnish rivers evaluated using benthic diatom communities. Journal of Applied Phycology. 2002;**14**:1-7

[74] Delta Environmental. Test method: Generic diatom indices. 2007. Available from: http://www.deltaenvironmental.com.au/management/Lab_methods/Generic_indices.htm [28 Nov. 2015]

[75] Lobo EA, Callegaro VLM, Hermany G, Bes D, Wetzel CA, Olivera MA. Use of epilithic diatoms as bioindicators from lotic systems in southern Brazil, with special emphasis on eutrophication. Acta Limnologica Brasieliensis. 2004;**16**(1):25-40

[76] Dela-cruz J, Pritchard T, Gordon G, Ajani P. The use of periphytic diatoms as a means of assessing impacts of point source inorganic nutrient pollution in South-Eastern Australia. Freshwater Biology. 2006;**51**:951-972

[77] Suphan S, Peerapornpisal Y, Underwood GC. Benthic diatoms of Mekong River and its tributaries in northern and North-Eastern Thailand and their applications to water quality monitoring. Maejo International Journal of Science and Technology. 2012;**6**:28-46

[78] Cocquyt C. Diatoms form a hot spring in Lake Tanganyika. Nova Hedwigia. 1999;**68**(3):425-439

[79] Ali AD, Ezra AG, Abdul SD. Species composition and distribution of freshwater diatoms from Upper Dilimi River, Jos, Nigeria. IOSR Journal of Pharmacy and Biological Sciences (IOSR-JPBS). 2015;**10**(5):53-60

[80] Compere P. Some algae for the Red Sea Hills in north-eastern Sudan. In: Dumont HJ, El Moghraby AI, Desougi LA, editors. Limnology and Marine Biology in the South Sudan, Vol. 21. Switzerland: Developments in Hydrobiology; 2012. 374 pp

[81] Kelly MG, Bennion H, Cox EJ, Goldsmith B, Jamieson J, Juggins S, et al. Common Freshwater Diatoms of Britain and Ireland: An Interactive Key (Online). Bristol: Environment Agency; 2005. Available from: http://craticula.ncl.ac.uk/EADiatomKey/html/taxon13810180.html [20 March 2016]

[82] Guiry MD, Guiry GM. Caloneis Bacillum (Grunow) Cleve 1894. Algae Base World-Wide Electronic Publication. Galway: National University of Ireland; 2016. Available from: http://www.algaebase.org/search/species/detail/?species_id=y1b89c1c56b056c40 [30 April 2016]

[83] Prygiel J, Coste M, Bukowska J. Review of the major diatom-based techniques for the quality assessment of rivers—State of the art in Europe. In: Prygiel J, Whitton BA, Bukowska J, editors. Use of Algae for Monitoring Rivers III. Douai: Agence de l'Eau Artois-Picardie; 1999. pp. 138-144

[84] Gomà J, Ortiz R, Cambra J, Ector L. Water quality evaluation in Catalonian Mediterranean rivers using epilithic diatoms as bioindicators. VIE MILIEU. 2004;**54**(2-3):81-90

[85] Naumann, E. 1921. Einige Grundlinien der regionalen Limnologie. Lunds University: Arsskr. N.F., 17(8)

[86] Reynolds CS. The Ecology of Freshwater Phytoplankton. United Kingdom: Cambridge University Press; 1984. 384 pp

[87] Gbadebo AM, Taiwo AM, Adeola AJ. Assessment of dissolved silica content of groundwater from southwestern Nigeria. Journal of Human Ecology. 2013;**43**(3):273-279

[88] Tan X, Zhang Q, Burford MA, Sheldon F, Bunn SE. Benthic diatom based indices for water quality assessment in two subtropical streams front. Microbiology. 2017;**8**:601. DOI: 10.3389/fmicb.2017.00601

[89] Vilbaste S. Application of diatom indices in the evaluation of the water quality in Estonian streams. Proceedings of the Estonian Academy of Sciences, Biology, Ecology. 2003;**53**(1):37-51

[90] Eassa AM. The use of diatom indices for the assessment of Shatt AL–Arab river water quality. Journal of Basrah Researches. 2012;**38**(1):114-124

[91] Solak CN, Ács É, Dayioglu H. The application of diatom indices in the Felent Creek (Porsuk-Kütahya). In: Van Dam, Cadee, editors. Proceedings of the Joint Meeting of the Dutch-Flemish Society of Diatomists (NVKD) and 3rd Central European Diatom Meeting (CE-DiatoM) 26-29 March, 2009. Diatomededelingen 33. 2009. pp. 107-109

Chapter 4

Functional Analysis of Stream Macroinvertebrates

Kenneth W. Cummins

Additional information is available at the end of the chapter

http://dx.doi.org/10.5772/intechopen.79913

Abstract

The worldwide study of stream ecosystems remains a topic of great interest, impacting methods and concepts critical to the preservation and management of global freshwater resources. Stream macroinvertebrates, especially aquatic insects, have served as one of the main pillars of inquiry into the structure and function of running water ecosystems. Stream macroinvertebrates have been used so extensively for over 100 years because they are universally present and abundant, can be readily observed with the unaided eye, (unlike algae and microbes) and are much less mobile than fish which can easily move to totally new locations. Although taxonomic identification has been the basis of analysis of stream macroinvertebrates, functional analysis now offers an additional tool that allows much more rapid analysis that can be accomplished in the field using simpler methodology.

Keywords: streams, macroinvertebrates, functional feeding groups, foods of stream invertebrates, surrogates for ecosystem attributes

1. Introduction

In 1970, Robert Pennak, the preeminent freshwater invertebrate biologist, held that the basic unit of all stream ecology studies should be species level taxonomy (personal communication). This view was shared by essentially all stream ecologists of the day. Given the condition of many stream ecosystems and the taxonomy of aquatic insects then and now [1] that was, and is, a severe impediment to the advancement of research on streams. An alternative approach, based on macroinvertebrate functional analysis, coupled with higher order taxonomy (family or, if possible, genus) was proposed to facilitate addressing stream ecosystem research questions [2, 3]. This functional analysis focuses on adaptations used by freshwater macroinvertebrates to acquire their food. In this approach, seven functional feeding groups

Functional feeding groups (FFG)	Examples of taxa	Adaptations for acquiring food resources
Scrapers (SC)	Ephemeroptera: Heptageniidae, Ephemerellidae *Drunella* Trichoptera: Uenoidae, Glossosomatidae Helicopsychidae, Psychomyiidae Hemiptera: Corixidae Coleoptera (larvae): Psephenidae, Elmidae Gastropoda	Mandibles with knife-like leading edge in aquatic insects, and file-like radula in Mollusca that removes attached algae; in Ephemeroptera, alga removal may be assisted by front legs
Algal piercers (APR)	Trichoptera: Hydroptilidae	Piercing mouth parts that suck contents from individual algal cells
Detrital shredders (DSH)	Plecoptera: Pteronarcyidae, Nemouridae, Capniidae, Peltoperlidae, Leuctridae, Taeniopterygidae Trichoptera: Limnephilidae, Calamoceratidae, Lepidostomatidae Tipulidae: *Tipula* Crustacea: Amphipoda, Isopoda, Decapoda	Chewing mouthparts, selection for softest portions of conditioned (colonized by microbes, especially aquatic hyphomycete fungi) vascular plant tissue
Gathering collectors (GC)	Ephemeroptera: Baetidae, Leptophlebiidae, Ephemerellidae, Tricorythidae, Caenidae Trichoptera: Leptoceridae, Odontoceridae Coleoptera: Elmidae (larvae), Hydrophilidae (adults) Diptera: Chironomidae Chironomini, Orthocladiinae Oligochaeta	Non-specialized mouth part morphology that facilitates sweeping fine FPOM into the mouth
Filtering collectors (FC)	Ephemeroptera: Isonychiidae Trichoptera: Hydropsychidae, Philopotamidae, Polycentropidae Diptera: Simuliidae, Chironomidae, Tanytarsini Mollusca: Sphaeriidae, Unionidae	Filtering fans or setae on front legs or silk nets or strands that trap FPOM from the passing water column

Functional feeding groups (FFG)	Examples of taxa	Adaptations for acquiring food resources
Herbivore shredders (HSH)	Lepidoptera: Crambidae, Noctuidae Coleoptera: Cocinellidae *Galerocella*	Chewing moth parts and crochets (Lepidoptera) that hold plant in place while feeding
Predators (P)	Plecoptera: Perlidae, Perlodidae Trichoptera: Rhyacophilidae Odonata: Anisoptera, Zygoptera Megaloptera: Corydalidae, Sialidae Hemiptera: Belastomatidae, Naucoridae, Coleoptera: Dytiscidae, Hydrophilidae (larvae), Dytiscidae (adults) Diptera: Tipulidae, Tabanidae, Empididae, Chironomidae, Tanypodinae	Crushing, piercing or grasping moth parts and/or front legs; active, with large eyes or ambush predators; with swimming hind legs, crawling legs or welts or prolegs

Categories based on morphological and behavioral adaptations for acquiring specific food categories [1, 3–5].

Table 1. Macroinvertebrate functional feeding groups (FFG).

(FFG) usually are coupled with their seven food categories. The relative abundance of the food categories matches with the relative abundance of the FFGs that utilize those food categories (**Table 1**) [1, 3–6].

Therefore, by identifying a limited number of food categories supporting stream macroinvertebrates it is possible to arrive at the morphological and behavioral adaptations generally shared by groups of taxa (FFG) that are adapted to acquire each of the food resource categories [7].

2. Functional groups (FFG) and food categories

Stream macroinvertebrate FFGs are listed below and summarized in **Table 1**.

1. Scrapers (SC) have morphological-behavioral adaptations that enable them to scrape non-filamentous attached algae from substrates (coarse sediments, wood, or stems of rooted aquatic vascular plants) in streams or lake littoral zones.
2. Detrital shredders (DSH) are adapted to feed on terrestrial plant litter (coarse particulate organic matter (CPOM), primarily leaves or needles that have been entrained in the stream and conditioned (colonized by) microbes, especially aquatic hyphomycete fungi.
3. Gathering collectors (GC) have very generalized adaptations used to feed on fine particulate organic matter (FPOM) of particle size less than 1 mm FPOM) which they sweep up

from depositional areas or crevices in flowing or turbulent where it has settled or been entrained.

4. Filtering collectors (FC) have adaptations that allow them to capture FPOM from the passing water in streams or resuspension by turbulence in lakes using morphological structures or silk capture nets.

5. Herbivore shredders (HSH) are adapted to feed on live rooted aquatic plants, primarily the leaves.

6. Herbivore piercers (HPCR) are adapted to pierce individual filamentous algal cells and suck out the cell contents (primarily Trichoptera, Hydroptilidae).

7. Predators (P) are adapted to catch and consume live prey by engulfing the prey or piercing and extracting the prey hemolymph.

Most genera of North American aquatic insects have been assigned to FFG categories in tables that appear after each taxonomic chapter in [1].

Parallel or convergent evolution has endowed differing taxonomic groups with similar morphological (e. g. mandible structure) and behavioral (e. g. net spinning and case construction) adaptations for acquiring a given food resource.

An example is the similar mandible structure found in larvae of four different scrapers (SC); three caddisfly (Trichoptera) genera representing three different families and one beetle

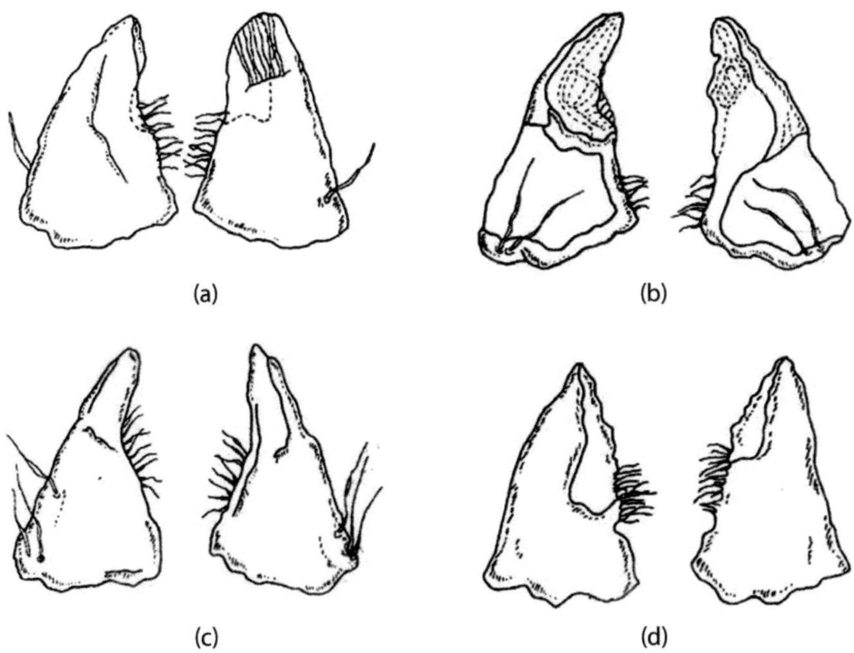

Figure 1. Mandible structure of algal scrapers (SC) in four different taxa: Three Trichoptera genera: (A) *Glossosoma*, (B) *Helicopsyche*, (C) *Neophylax* in different families and a Coleoptera genus (D) *Psephenus*. All have scraper leading edges and basal setae to aid in retention and passage into the mouth of the removed algae. Modified from [8].

(Coleoptera) genus, *Psephenus*, in the Family Elmidae are shown in **Figure 1** [8]. All the mandibles have a knife-like leading edge which, when drawn across a rock or wood surface towards the mouth and remove attached algae. This material is contained and directed into the mouth by basil setae **Figure 1**. A further example is found in the mandibles of wood gouging detrital shredders (HSH) shown in **Figure 2** [9]. Three different genera in three different families and orders are represented. The mandibles of these wood shredder aquatic insects are all three-toothed, scooped-shaped with basal setae. The mandibles are used to gouge grooves in the surface of wood entrained in a stream, ingest it and digest the microbes present that are the source of their nutrition. This poor quality food resource results in longer

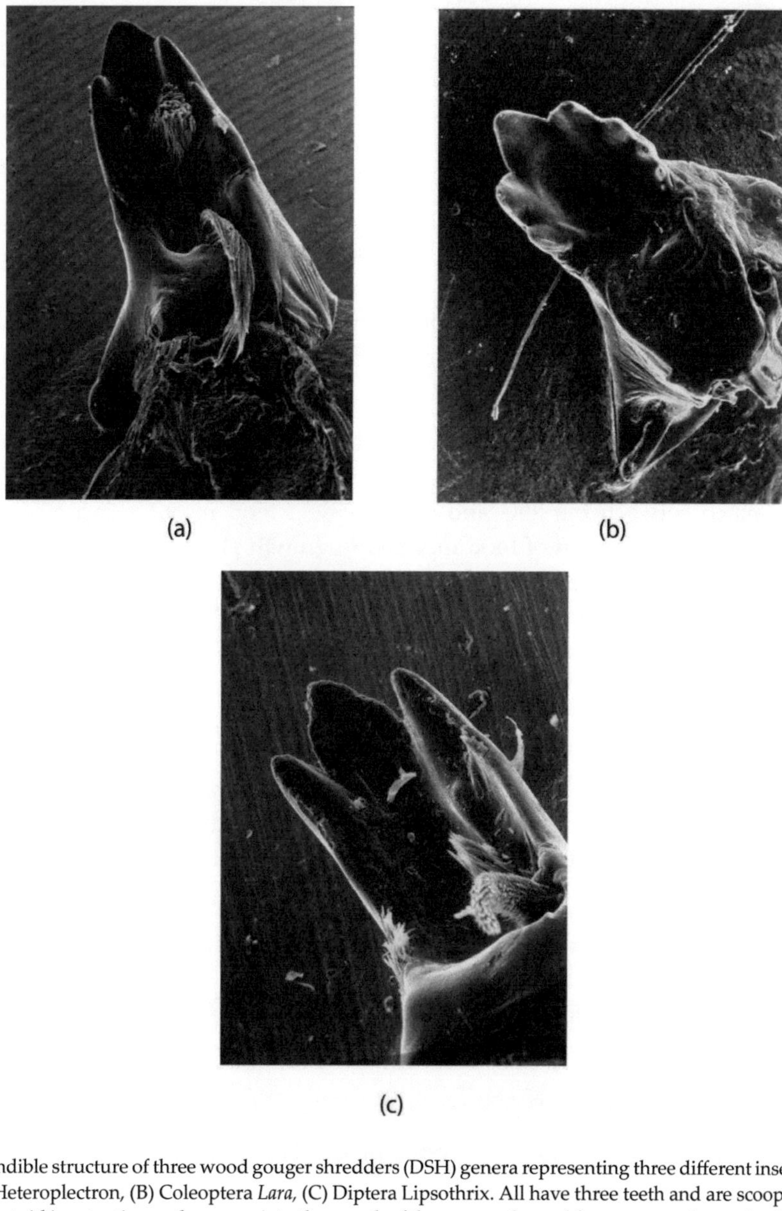

Figure 2. Mandible structure of three wood gouger shredders (DSH) genera representing three different insect orders. (A) Trichoptera Heteroplectron, (B) Coleoptera *Lara,* (C) Diptera Lipsothrix. All have three teeth and are scoop-shaped with basal setae that aid in retention and passage into the mouth of the removed wood fragments and contained microbes [9].

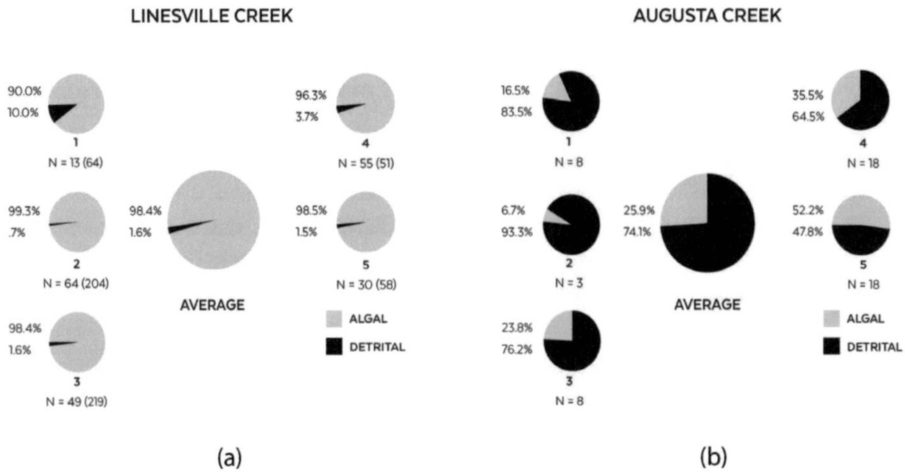

Figure 3. Gut analyses of each of the five instars, and the overall average of larvae of the Trichoptera species *Glossosoma nigrior* form two different streams. (A) Linesville creek, a second order stream in Pennsylvania, USA, with diatom dominated periphyton (average of all five instars = 98.4% algae) and (B) Augusta creek, a second order stream in Michigan, USA, with a detritus dominated periphyton (average of all five instars = 74.1% detritus) [9].

life cycles, especially the larvae of the beetle *Lara* that requires 5 years to mature. Often, *Lara* is found on long term stable large wood habitat in northwest, old growth conifer bordered streams [10, 11].

Many FFGs are restricted to a single mode of feeding. These obligate taxa maintain the same FFG mode of feeding, independent of the quality of the food being harvested [12]. A measure of the match between the FFG and the harvested food resource can be described by the efficiency of the conversion of food ingested to growth [4]. An example is the caddisfly *Glossosoma nigrior* (Trichoptera Glossosomatidae) [12]. Analysis of gut contents of two populations of *G. nigrior* occurring in two different streams, where they both fed as FFG scrapers, revealed that the ingested food differed significantly through all five instars of the growth period (**Figure 3**) [9]. The *G. nigrior* population in the stream that provided good non-filamentous algal periphyton had gut contents totally dominated by algae. The pre-pupae in this population achieved significantly higher final dry biomass than the other population which had gut contents dominated by detritus. This second stream was in a canopy closed forest in which the rock surfaces were covered with fine detritus with little or no algal periphyton. The gut analysis method involving the use of Millipore™ filters is described in [13].

Facultative FFG taxa are not as fixed with regard to adaptations for acquiring food. The added flexibility in feeding allows for survival in a wider range of habitats that offer more food types but the conversion of ingested food to growth is less efficient than in obligate taxa. In their early instars, the majority of taxa are facultative generalists feeding on detritus, even predators like the megalopteran *Nigronia* [14].

A picture key to stream macroinvertebrate FFGs appears as an Appendix in [5].

Stream macroinvertebrate FFG food categories are listed below and described in **Table 2** [4, 5].

Functional feeding groups (FFG)	Foods and descriptions	Typical habitats
Scrapers (SC)	Periphyton: single cells or colonies of non-filamentous algae, especially diatoms, attached to the substrate with loosely associated FPOM, microbes, and/or micro-arthropods	Stream riffles and runs, lake wave-swept shore lines, or rooted vascular aquatic plant beds
Algal piercers (APR)	Filamentous algae: green, blue green, or red algae	Riffles, pools, stream margins or back waters where ever filamentous algae occurs.
Detrital shredders (DSH)	CPOM: riparian vascular plant litter conditioned in the water (colonized by aquatic hyphomycete fungi and bacteria). Micro-arthropods may be present	Plant litter accumulations against obstructions in the current (leaf packs) or settled in pools and on the bottom in lake plant beds
Gathering collectors (GC)	FPOM: organic particles surface colonized by bacteria or mineral particles with organic coating	Pools and other depositional areas in streams and lakes
Filtering collectors (FC)	FPOM: same as above	Particles suspended in the current or by wave action
Herbivore shredders (HSH)	Plant beds: live rooted aquatic vascular plants	Any stream or lake habitats where plant beds grow
Predators (P)	Prey: live aquatic invertebrates in the size range that can be captured by a predator	Essentially all stream and lake habitats where prey are found

After [4, 5]. FPOM is fine particulate organic matter of particle size <1 mm. CPOM is coarse particulate organic matter of particle size >1 mm.

Table 2. Macroinvertebrate functional feeding groups (FFG) food categories they are adapted to acquire and typical habitats where the food resources will be found.

1. Attached non-filamentous algae, primarily diatoms but also green or red algae.
2. Terrestrial plant litter (CP)M) entrained in the stream and colonized and conditioned by microbes.
3. Depositional FPOM on fine sediments in pools, stream margins or under or in crevices in coarse sediments in the current.
4. Suspended FPOM transported in the water column or suspended by turbulence.
5. Living rooted vascular aquatic plants.
6. Filamentous green algae.
7. Live prey.

3. Similar FFG adaptations in different taxa

Noel Hynes, arguably the premier stream ecologist of the last 60 years, observed that he could "turn over a rock in any clean stream the world over and recognize familiar aquatic

Figure 4. Two scraper (SC) mayfly nymphs from two different families from two continents. Ephemeroptera: (A) Heptageniidae from North America and (B) Ephemeroptera Leptophlebiidae from Brazil. Both nymphs are dorsoventrally flattened with the eyes and antennae located dorsally and the scraper mouth parts ventrally positioned on the head including the scraper mouth parts [9, 15].

insects, but they were all in different families" (personal communication). An example is the very similar dorsal-ventrally flattened body form of a North American scraper mayfly (Ephemeroptera, Heptageniidae) and a Brazilian mayfly (Ephemeroptera, Leptophlebiidae) (**Figure 4**) [9, 15]. Both these scrapers primarily use the mandibles to remove attached periphytic algae from rock surfaces. The mandibles of different taxa functioning as scrapers are similar: a cutting ventral edge with basal setae that keep the algae confined as it is moved into the mouth (**Figure 1**). Another example of similar feeding structures was described above for wood-gouging shredders belonging to different taxa (**Figure 2**). All these examples represent parallel or convergent evolution resulting in differing taxa being adapted to acquire the same food resource type.

The FFG of a given taxon can vary with age (instar in aquatic insects). There is evidence that essentially all aquatic insects are gathering collectors in the first (and second in some) instars. For example, Petersen [14] documented through gut analysis that the first instar of the predaceous megalopteran *Nigronia serricornis* (Corydalidae) fed on FPOM as a gathering collector. Taxa can also switch FFG as they progress through the growth period. The limnephilid caddisfly (Trichoptera) *Pycnopsyche lepida* switches from feeding on conditioned CPOM leaf litter as a detrital shredder in the first four instars to a scraper in the fifth (final) instar. This transition is readily discernable because in the first four instars the larvae construct an organic case made of leaf sections and fine sticks that are readily available in the litter accumulations where the larvae occur. In the fifth instar the larvae move into fast water where the feed by scraping periphytic algae from cobbles in flowing water. At that time, the larvae convert to heavier mineral cases [15, 16].

4. Benefits of FFG analysis

There are two significant benefits in the FFG approach. First, it allows stream macroinvertebrates collected live in benthic or drift samples to be placed in ecologically relevant categories using only the level of taxonomy needed to separate them [6]. Simple examples would be: all Odonata dragonfly and damselfly nymphs are predators (P) and stone case-bearing Trichoptera (caddisfly) larvae are scrapers (SC) while those in organic cases (leaf and stick/stem material) are detrital shredders (DSH). These examples, in which ordinal taxonomy is all that is required, are used in the key given in Merritt et al. [5]. Clearly, the more taxonomic resolution the better, but the FFG allows typical volunteer or class groups to gather useful data on aquatic ecosystems of local concern. And, importantly, the data can be taken on site using live specimens that allow behavioral and color paten traits, that are lost or fade in preservation, to be used in the evaluation. Furthermore, the collected organisms can be preserved and returned to a laboratory, if accessible, when detailed taxonomic determinations are required, for example to calculate diversity indices (e. g. see ecological tables in [1]).

The FFG detrital shredders (DSH) are reliably linked to the inputs plant litter from the riparian zone [17] that can be studied easily with the use leaf or needle packs which accurately mimic how plant litter is naturally processed by fungi and detrital shredders (DHS) in streams [18]. These are prepared using leaves or needles of trees that are present in the riparian zone of a stream under study (**Figure 5**) [8, 18]. Dry leaves or needles are weighed into 10 g amounts (to nearest 0.01 g) and soaked in warm water until soft (5–10 minutes). When the leaves are soft enough to handle so that they do no break, they are gathered into a packs and stapled together with the plastic Tees using a Buttoneer™ gun. The needles are threaded into a chain on monofilament fishing line (**Figures 5** and **6**). Each leaf or needle pack is fastened to an elastic band of sufficient size so that it can be attached to a common brick (**Figure 6**). The packs, one or two to a brick (**Figure 6**), are placed in the stream facing into the current to simulate the obstructions against which plant litter accumulates in streams in the current. It is important that some flow occurs across the surface of the leaves or needles to maintain dissolved oxygen levels required by the obligate aerobe aquatic Hyphomycete fungi that colonize the plant material and constitute the major source of the nutrition for detrital shredders (DHS) [19]. The leaf pack shown in **Figure 6** are hickory leaves that were incubated in a third order woodland Michigan stream in October for 2 weeks at 10°C. The effect of the shredder feeding is evident. The softer, most heavily fungal colonized portions of the leaf have been used [20, 21]. The detrital shredders (DSH) select portions of the leaves or needles that are most heavily colonized by aquatic hyphomycete fungi. Specific polyunsaturated lipids of the fungi attract the shredders [19].

(a) (b)

Figure 5. Leaf packs stapled together with plastic Ts using a Buttoneer™ gun and fastened to an elastic band and attached to a common brick prior to placement in a stream (A). The conifer needles are strung on fishing monofilament line prior to fastening to an elastic band and attachment to a common brick for placement in a stream (B). Such leaf and needle packs can be used as a bioassay for microbial-shredder (DSH) processing of CPOM in a stream ecosystem.

The second benefit of using the FFG approach is that ratios of the number of specimens collected in each FFG category can be used to describe a stream reach and compare it to other reaches and other streams (**Table 3**) [5, 8]. Because these ratios are dimensionless numbers, they are relatively independent of sample size. For example, it has been demonstrated that the *relative* number of scrapers collected from one rock in a given stream riffle is not statistically different from the collection from five rocks in the same riffle [9]. Some FFG ratios that can serve as surrogates for stream ecosystem attributes are summarized in **Table 4**. Threshold ratio values (percentages) have been proposed based on field evaluations that can serve as

Figure 6. Deciduous leaf pack after 2 weeks at 10°C in a woodland stream in Michigan. The effect of detrital shredder (DSH) feeding s evident. The softer leaf tissues with the greater biomass of aquatic Hyphomycete fungi have been consumed before the more lignified leaf veins. The plastic Buttoneer™ Tees that held the leaves together are visible.

Stream ecosystem attributes	Name of FFG surrogate ratio	Definition of FFG surrogate ratio
Autotrophic vs. heterotrophic energetics	P/R: autotrophy to heterotrophy index	Gross primary production compared total community respiration (primary production/respiration or P/R)
CPOM vs. FPOM	Shredder index	CPOM riparian plant litter compared to riparian FPOM + FPOM generated within the stream (e. g. macroinvertebrate feces, mechanical fragmentation microbes, DOM flocculation)
Suspended vs. storage FPOM	Filtering collector index	Suspended FPOM transported in the current compared to FPOM in storage (entrained) in or on the sediments
Stable vs. unstable sediments	Habitat stability index	Coarse sediments + large wood + bed rock + rooted vascular plants compared to small easily moved sediments + FPOM
Top down vs. bottom up macroinvertebrate communities	Predator top down index	Predator regulation of macroinvertebrate communities as compared to regulation by in stream primary production + detritus support of macroinvertebrate communities

Based on [5, 8]. FPOM is fine particulate organic matter of particle size <1 mm, CPOM is coarse particulate organic matter of particle size >1 mm, DOM is dissolved organic matter.

Table 3. Use of functional feeding groups (FFG) as surrogates for stream ecosystem attributes.

Ratio name	FFG surrogate ratios	Proposed thresholds and explanations	Interpretations
Autotrophy-heterotrophy index	SC + HSH + APC to DSH + GC + FC	FFG ratio of >0.75 corresponds to a directly measured P/R = 1.00. Represents in stream plant (algae and vascular) production > than riparian plant litter inputs (or total respiration of microbes + plants and animals)	Stream energetics driven by periphytic algal + any vascular plant production as compared to riparian plant litter inputs
Shredder index	DSH + HSH to GC + FC	FFG ratio of >0.50 in fall-winter or >0.25 in spring-summer. Represents CPOM shredder food availability > FPOM collector food availability	CPOM food support for shredders > than FPOM for Collectors. Fall-winter litter inputs usually >spring-summer and condition more rapidly
Filtering collector index	FC to GC	FFG ratio of >0.50 indicates suspended FPOM load > storage (entrained) FPOM	FPOM food for collectors at higher density and/or better quality than storage FPOM
Habitat stability index	FC + SC + HSH to DSH + GC	FFG ratio of >0.50 indicates that stable locations for scraping and attachment are in greater abundance than shifting unstable substrates	Filtering collectors require stable locations for attachment and construction of capture nets and scrapers require surfaces that remain in a stable position facing up
Top down predator index	P to total FFGs	Predator to prey ratio 0.10–0.20 to total macroinvertebrate population	This level of predator population density (or biomass) allows for sufficient prey to support them. If predators >20% probably indicates populations of rapid turnover (polyvoltine prey populations present

Proposed thresholds after [22–25]. SC = scrapers, HSH = herbivore shredders, DSH = detrital shredders, GC = gathering collectors, FC = filtering collectors, APC = algal piercers, P = predators.

Table 4. FFG surrogate ratios for stream ecosystem attributes, proposed surrogate ratio thresholds and resulting interpretations.

surrogates to ecosystem attributes such as the classification of a stream reach as autotrophic (i.e. dependent on in-stream primary production as the primary energy source) vs. heterotrophic (dependent on riparian out of steam primary production as the primary energy source) (**Table 4**) [22–25]. This particular ratio (SC + HSH/DSH + SC + FC) is a surrogate for directly measured P/R, which is the ratio of gross primary production to total community respiration (i. e. including autotrophs). P/R, which is measured by monitoring oxygen levels over time across a stream reach or in enclosed *in situ* chambers [26]. The FFG surrogate P/R ratio is strongly influenced by season and like other FFG ratios might require a different threshold for spring-summer vs. fall-winter (**Table 4**).

In the case of detrital shredders (DSH), their presence and abundance depends on the type of plant litter inputs and season. If the riparian zone is dominated by deciduous hard woods, the inputs are in the fall–winter period. Hardwoods, except oaks, are conditioned rapidly by aquatic hyphomycetes and fed on by shredders. Conifers (evergreens) shed foliage primarily in the spring–summer and conditioning by fungi is much slower and shredder feeding is delayed. Prairie Creek in Redwood National Park, California provides an example from an old growth conifer forest stream with slow fungal conditioning and delayed detrital shredder (DSH) activity (**Table 5** [27]). The riparian derived CPOM is largely wood and conifer needles which require along conditioning times before shredders begin feeding. This suggests that samples taken in December were too early to detect the major shredder activity. The shredder caddisfly larvae (Trichoptera: *Lepidostoma* and *Gumaga*) and stonefly Nymphs (Plecoptera: Peltoperlidae, Capniidae, Leuctridae) that were collected were quite small. The mean dry mass per individual, of the 34 individuals collected was 6.6 mg (caddisflies 111 mg, stoneflies 75 mg) indicating very early instars. The ratio was low because the biomass of the GC (0.061 mg) plus FC (3.35 mg) was much greater. The recommended threshold for conifer old growth forest streams should be based samples for February or March. The general 0.50 threshold was based on data from deciduous forest streams taken in the fall when the conditioning of the riparian litter was sufficient to accommodate expected shredder populations [18].

There is an extensive literature using FFG analyses, for example the numerous references cited in the ecological tables that accompany each aquatic insect order in [1]. FFG analyses and the ratio method, including proposed thresholds have been used to evaluate Florida Rivers. FFGs were used to characterize the un-dammed reach of the Kissimmee River as a model for general restoration of the 100 miles of the channelized River [22]. The ratio method was employed to characterize the remnant oxbows of the Caloosahatchee River according to their ecological attributes and to provide recommendations for preservation and restoration of the River's oxbows [25]. Floodplain (marsh) habitats of the St. John's River were also evaluated relative to hydrological influences using FFG surrogate ratios to predict the effects of water withdrawal from the river [25]. The method also was used to characterize the ecological conditions in a wide range of Brazilian streams and rivers [24].

Stream FFG ratios	Calculated ratios		Proposed thresholds	Stream ecosystem interpretations
	Numbers	Dry biomass (mg)		
P/R index: (SC + HSH/DSH = GC = FC)	0.96	0.55	Ratio > 0.75 (autotrophic)	SC numbers indicate a significant algal periphyton based autotrophic stream ecosystem (significant biomass of stonefly and caddisfly shredders indicate heterotrophic system supported by riparian plant litter inputs)
CPOM shredder index: (DSH/ GC = FC)	0.15	0.04	Ratio > 0.50 (predicted fall–winter shredder component)	The predicted range for fall–winter shredder populations not met. (supply of sufficiently conditioned [colonized by microbes] riparian plant litter inputs, not adequate)

Stream FFG ratios	Calculated ratios		Proposed thresholds	Stream ecosystem interpretations
	Numbers	Dry biomass (mg)		
FPOM suspended load index: (FC/ DSH + GC + FC)	0.14	0.28	Ratio > 0.50 (sufficient FPOM to support filtering collectors, assuming sufficient quality)	The amount, or quality (e.g.. organic content) of FPOM in suspension and transport below typical levels to support FC populations (this old growth conifer forest stream dominated by slowly processed wood and litter likely generated only primarily low amounts of poor quality FPOM; rapid turnover deciduous litter from big leaf and vine maple probably absent)
Channel stability index: (SC + FC + HSH/DSH + GC)	1.09	0.96	Ratio > 0.50 (sufficient stability to support SC and FC populations}	Stable substrates, coarse sediments and large wood, well above levels predicted to support surfaces required scraping and filtering macroinvertebrates (large wood in old growth forest streams observed to provide a long term stable habitat)
Top down predator index: (P % of total)	0.15	0.21	Ratio > 0.10 < 0.20 (predicted predator top down control of macroinvertebrate populations)	Both numerical and biomass estimates of predators fall within the expected % r of the macroinvertebrate community (all the expected predators present; stoneflies, especially the large Perlidae, Tipulidae [except *Tipula*], *and Tanypodinae midges*)

Modified from [27].

Table 5. FFG ratio analysis of macroinvertebrate benthic summer samples taken in Prairie Creek winter (December).

5. Conclusions

Proposed in this chapter is the use of the FFG method of analysis to gain rapid and efficient insight into macroinvertebrate community composition and fits unction in freshwater ecosystems. The method should be compatible with a broad level of expertise, from beginner to advanced, and can be conducted stream-side allowing live animals to be released or preserved and returned to the lab. If the FFG ratio method is to be used, at least some qualitative observational data should be recorded: riparian information about the dominant vegetation cover (e. g. percent deciduous vs. evergreen) and the % canopy closure; dominant stream habitat (riffles vs. pools, coarse vs. fine sediments, % accumulation plant litter such as leaf packs); discharge level relative to bank full (the permanent vegetation line)and general land use observations (e. g. agricultural or live stock grazing, timber harvest or other human disturbances). When conducting FFG analysis it is most useful to collect, and keep separate, samples from three types of habitat: Riffles (coarse sediments), pools (fine sediments), and plant litter accumulations. Samples can be taken with timed (e. g. 15 s) D-frame or aquarium net samples. The data can be combined later into a "composite" sample, but the relative importance of each habitat can be assigned based on the qualitative evaluation of the % stream bottom cover of each habitat type. Most importantly, The FFG analysis technique does not foreclose on any traditional use of taxonomic analysis of the samples in the laboratory.

Acknowledgements

Thinking back to 1973, I first became fascinated with adaptations of freshwater macroinvertebrates, especially in streams. Herb Ross Illinois Natural History Survey) invited me to submit a contribution to the Annual Review of Entomology. He said there were rarely any papers on aquatics and suggested I consider ecological adaptations of aquatic insects as a topic. At about the same time, Noel Hynes (University of Waterloo, Canada) told me he could turn over a rock in any stream in the world and recognize the mayfly nymphs, but they were all in the "wrong family." This example of convergent or parallel evolution formed the basis for the functional feeding group approach. During the intervening years, a monumental array of colleagues has been instrumental in my understanding of freshwater macroinvertebrates. George Lauff (at the University of Michigan) directed my PhD work and Rich Merritt (Michigan State University) and Peggy Wilzbach (Humboldt State University, CA) have provided ideas and encouragement for decades.

Author details

Kenneth W. Cummins

Address all correspondence to: kc8161@gmail.com

Fisheries Biology, Humboldt State University, Arcata, CA, USA

References

[1] Merritt RW, Cummins KW, Berg MB. An Introduction to the Aquatic Insects of North America. 4th ed. Dubuque, IA: Kendall/Hunt; 2008. 1158 p

[2] Cummins KW. Trophic relations of aquatic insects. Annual Review of Entomology. 1973;**18**:183-206

[3] Cummins KW. Structure and function of stream ecosystems. Bioscience. 1974;**24**:631-664

[4] Cummins KW, Klug MJ. Feeding ecology of stream invertebrates. Annual Review of Ecology and Systematics. 1979;**10**:147-172

[5] Merritt RW, Cummins KW, Berg MB. Trophic relationships of macroinvertebrates. Chapter 20. In: Hauer FR, Lamberti GA, editors. Methods in Stream Ecology. London: Academic Press (Elsevier); 2017. pp. 413-433

[6] Cummins KW. Combining taxonomy and function in the study of stream macroinvertebrates. Journal of Limnology. 2016;**75**:235-241

[7] Cummins KW, Wilzbach MA. Rivers and streams: Ecosystem dynamics and integrating paradigms. In: Encyclopedia of Ecology. London, UK: Academic Press, Elsevier; 2008. pp. 3084-3095

[8] Cummins KW, Merritt RW, Berg MB. Ecology and distribution of aquatic insects. Chapter 6. In: Merritt RW, Cummins KW, Berg MB, editors. An Introduction to the Aquatic Insects of North America. 4th ed. Dubuque, IA: Kendal/Hunt; 2008. pp. 105-122. 1158 p

[9] Cummins KW. Unpublished

[10] Anderson NH, Sedell JR, Roberts M, Triska FJ. The role of aquatic invertebrates in processing wood debris in coniferous forest streams. American Midland Naturalist. 1978;**100**:64-82

[11] Anderson NH, Steedman RJ, Dudley T. Patterns of exploitation by stream invertebrates on wood debris (xylophagy). Vereiratet. Internationale Vereingte Limnology. 1984, 1984;**22**:1847-1852

[12] Anderson NH, Cummins KW. The influence of diet on the life histories of aquatic insects. Journal of the Fisheries Research Board of Canada. 1979;**36**:335-342

[13] Coffman WP, Cummins KW, Wuycheck JC. Energy flow in a woodland stream ecosystem. 1: Tissue support trophic structure of the autumnal community. Archiv fur Hydrobiologie. 1971;**68**:232-276

[14] Petersen RC. Life history and bionomics of *Nigronia serricornis* (Say) (Megaloptera: Corydalidae) [PhD thesis]. East Lansing, Michigan: Michigan State University; 1974. 210 p

[15] Hamada N, Nessimian JL, Querino IB. Beotropical Insetos aquaticos na Amazonia: Taxonomia, biologia, ecologia. Manaus, Brazil: Editora do INPA; 2014. 724 p

[16] Cummins KW. Factors limiting the microdistribution of the caddisflies *Pycnopsyche lepida* and *Pycnopsyche guttifer* in a Michigan stream (Trichoptera: Limnephilidae). Ecological Monographs. 1964;**34**:271-295

[17] Grubbs SA, Cummins KW. Processing and macroinvertebrate colonization of black cheery (*Prunus serratona*) in two streams differing in summer biota. American Midland Naturalist. 1994;**132**:284-293

[18] Cummins KW, Wilzbach MA, Gates DM, Perry JB, Taliferro WB. Shredders and riparian vegetation. Bioscience. 1989;**39**:24-30

[19] Hanson BJ, Cummins KW, Cargill AS, Lowery RR. Lipid content, fatty acid composition, and the effect of diet on fats of aquatic insects. Comparative Biochemistry and Physiology. 1985;**808**:257-276

[20] Petersen RC, Cummins KW. Leaf processing in a woodland stream. Freshwater Biology. 1974;**4**:343-368

[21] Grubbs SA, Cummins KW. A leaf toughness method for directly measuring the processing of naturally entrained leaf detritus in streams. Journal of the North American Benthological Society. 1994;**13**:68-77

[22] Cummins KW. Evaluation of Kissimmee River in South Florida. In: Hamilton SW, White DS, Chester EW, Finley MT, editors. Proceedings of the 8th Symposium; Natural History of the Lower Tennessee and Cumberland River Valleys. Clarksville, Tennessee: Austin Peay State University. The Center for Field Biology; 1999. pp. 1-20

[23] Merritt RW, Cummins KW, Berg MB, Novak JA, Higgins MJ, Wessell KJ, et al. Development and application of a macroinvertebrate functional group approach in the bioassessment of remnant oxbows in the Caloosahatchee River in Southwest Florida. Journal of the North American Benthological Society. 2002;**21**:290-310

[24] Cummins KW, Merritt RW, Andrade P. The use of invertebrate functional groups to characterize ecosystem attributes in selected streams and rivers in Southeast Brazil. Studies on Neotropical Fauna and Environment. 2005;**40**:69-89

[25] Mattson RA, Cummins KW, Merritt RW, McIntosh M, Campbell E, Berg MB, et al. Hydroecological monitoring of benthic invertebrate communities of marsh habitats in the upper and middle St, Johns River. Florida Scientist. **77**:144-161

[26] Wilzbach MA, Cummins KW. Rivers and streams: Physical setting and adapted biota. In: Encyclopedia of Ecology. London, UK: Academic Press (Elsevier); 2008. pp. 3095-3106

[27] Cummins KW, Wilzbach MA, Kolouch B. Estimating dry biomass for freshwater macroinvertebrate functional groups and implications for stream ecosystem evaluations. Aquatic Insects. 2018; in press

Chapter 5

Monitoring Water Siltation Caused by Small-Scale Gold Mining in Amazonian Rivers Using Multi-Satellite Images

Felipe de Lucia Lobo,
Evlyn Márcia Leão de Moraes Novo,
Claudio Clemente Faria Barbosa and
Vitor Hugo Fernandes de Vasconcelos

Additional information is available at the end of the chapter

http://dx.doi.org/10.5772/intechopen.79725

Abstract

The small-scale mining techniques applied all over the Amazon river basin use water from streams, including digging and riverbed suctioning, rarely preventing environmental impacts or recovery of the impacted areas. As a consequence, thousands of tons of inorganic sediment (which can contain mercury) have been discharged directly into the rivers creating sediment plumes that travel hundreds of kilometers downstream with unknown consequences to the water quality and aquatic biota. We hypothesize that because of intensification of mining activities in the Brazilian Amazon, clear water rivers such as the Tapajós and Xingu rivers and its tributaries are becoming permanently turbid waters (so-called white waters in the Amazonian context). To investigate this hypothesis, satellite images have been used to monitor the sediment plume caused by gold mining in Amazonian rivers. Given the threat of intense water siltation of the Amazonian rivers combined with the technological capacity of detecting it from satellite images, the objective of this chapter is to inform the main activities carried out to develop a monitoring system for quantifying water siltation caused by small-scale gold mining (SSGM) in the Amazon rivers using multi-satellite data.

Keywords: total suspended solids, water quality change, remote sensing

1. Introduction

The Amazon river basin is the largest in the world, draining approximately 7,500,000 km^2 and discharging 20% of the global riverine waters into the ocean [1]. Within this territory, which includes eight countries in South America (Brazil, Bolivia, Colombia, Ecuador, Guyana, Peru, Suriname, and Venezuela), several economic activities are threatening water quality. The three activities with the most impact are hydropower dams, agribusiness expansion (deforestation) and, mostly, mining activities [2–5]. The political and economic context in Brazil is even more concerning, given the construction of several hydropower dams and the recent changes in environmental laws to benefit agribusiness and mining activities in the Amazon region. With regard to mining, although it usually takes place in small areas compared to cattle and soybean production, for example, the environmental impacts such as area degradation, water siltation, and metal contamination are more severe and intense than those of other land use changes [6].

In the Amazon, substantial small-scale gold mining (SSGM) activities started in the 1950s at a few sites, called "garimpos." In the 1980s, encouraged by the high gold price, hundreds of thousands of people migrated to mining sites, causing an intense "gold rush" in order to escape complete social downgrading [7]. The SSGM decreased in the 1990s due to overexploitation of superficial gold, gold price stagnation, and national economic crises. However, within recent years, the gold price has risen from US$ 400 per ounce (28.3 grams) in 2005 to US$ 1300 in 2016 encouraging a new gold rush in the Amazon region [8]. In Brazil alone, approximately 130,000 small-scale gold miners are responsible for 30 tons per year [9, 10]. This production generates at least R$ 26 million per month for the regional economy [11].

Despite its financial contribution, the semi-mechanized nature of small-scale gold mining activities often generates a legacy of extensive environmental degradation, both during operations and well after mining activities have ceased [7]. SSGM takes place mostly over alluvial deposits (river network) using either dredges or water-jet systems that cause dislodging of bottom or topsoil, respectively [11, 12]. The discharge of sediment into the water has severe impacts on the water quality, such as decreasing light availability for primary production [13] and changing benthic [14] and fish communities [15]. Currently, these socio-environmental impacts can be intensified since legislation is currently being debated in the Brazilian congress, which would weaken existing environmental/indigenous laws and to slash funding for environmental protection agencies.

Technically, monitoring of environmental impacts caused by SSGM, such as quantification of water siltation using satellite images, is rarely performed either due to lack of water quality data [e.g., total suspended solids (TSS)] or because of limitation of satellite image specifications such spatial resolution [16, 17]. Today, with the availability of 10 m images such as MSI/Sentinel-2, even narrow rivers (up to 30 m width) can be monitored.

2. Scientific hypothesis and objective

The scientific hypothesis of this research is that naturally low-turbid rivers (clear and black waters, based on the Sioli [18] and Junk [19] classification of the Amazonian water types) and some tributaries that are heavily impacted by SSGM tailing discharge are becoming "white" waters (naturally turbid waters), directly affecting the aquatic and benthic ecosystems.

To test the hypothesis of whether or not SSGM activities are responsible for the "whitening" process in some Amazonian rivers, we need to investigate the spatial-temporal TSS distribution along these rivers throughout the years.

Recently, Lobo et al. [20] have estimated TSS in the Tapajós river using an empirical model based on measured TSS and radiometric data. The effects of TSS derived from mining activities on both the inherent and apparent optical properties were quantified. The authors concluded that the inorganic nature of mine tailings is the main component affecting the underwater scalar irradiance in the Tapajós river basin. For tributaries with low or no influence of mine tailings, waters are relatively more absorbent. On the other hand, with TSS loadings from mining operations, the scattering process prevails over the absorption coefficient at the green and red wavelengths. This change in load and seasonality might affect, in the long run, biota composition of a previous clear water environment to a distinct light availability regime as the river becomes subjected to mining operations.

This approach worked well for the study area, and it seems a very promising approach for other areas in the Amazon region that are facing the same SSGM impacts in order to provide easy-access information for land use management in very remote areas with little financial support. However, further extension to other study areas in the Brazilian Amazon scale using different satellite data requires improvement on image processing to handle a large database and requires validation of the methods used for SSGM area mapping and TSS estimation prior to extend it to additional areas.

Therefore, the purpose of this chapter is to inform the main activities carried out toward a monitoring system for quantifying water siltation caused by SSGM in the Amazon rivers using multi-satellite data. In the assessment phase, the activities aimed to evaluate the viability of retrieving total suspended solids (TSS) from rivers other than Tapajós using images of several satellite missions.

3. Building a water siltation monitoring system

As a part of the assessment phase, a monitoring system was designed for areas selected according to the following criteria: (i) river basins that naturally present low sediment content (clear and blackwaters according to Junk [19], as opposed to white waters), in which discharge of mining tailings can have a negative impact on biodiversity and on the local economy; (ii) areas where SSGM is actively occurring over topsoils dominated by inorganic fine particles [21]; and (iii) rivers detectable by medium spatial resolution images or high resolution (at least 30 m wide). Considering these criteria, four areas were defined: (A) the Tapajós river, (B) the Amanã river, (C) Peixoto de Azevedo river, and (D) the Xingu river (**Figure 1**).

The first step was to select cloud-free images from the web platforms for each sensor. The images were downloaded from USGS, ESA, or DGI website servers (**Table 1**). In this chapter, few images are used here to illustrate the methodology. The second step was to convert top of the atmosphere (TOA) reflectance into surface reflectance (ρw). Considering the different sensors, the image correction may have different correction approaches (**Table 1**).

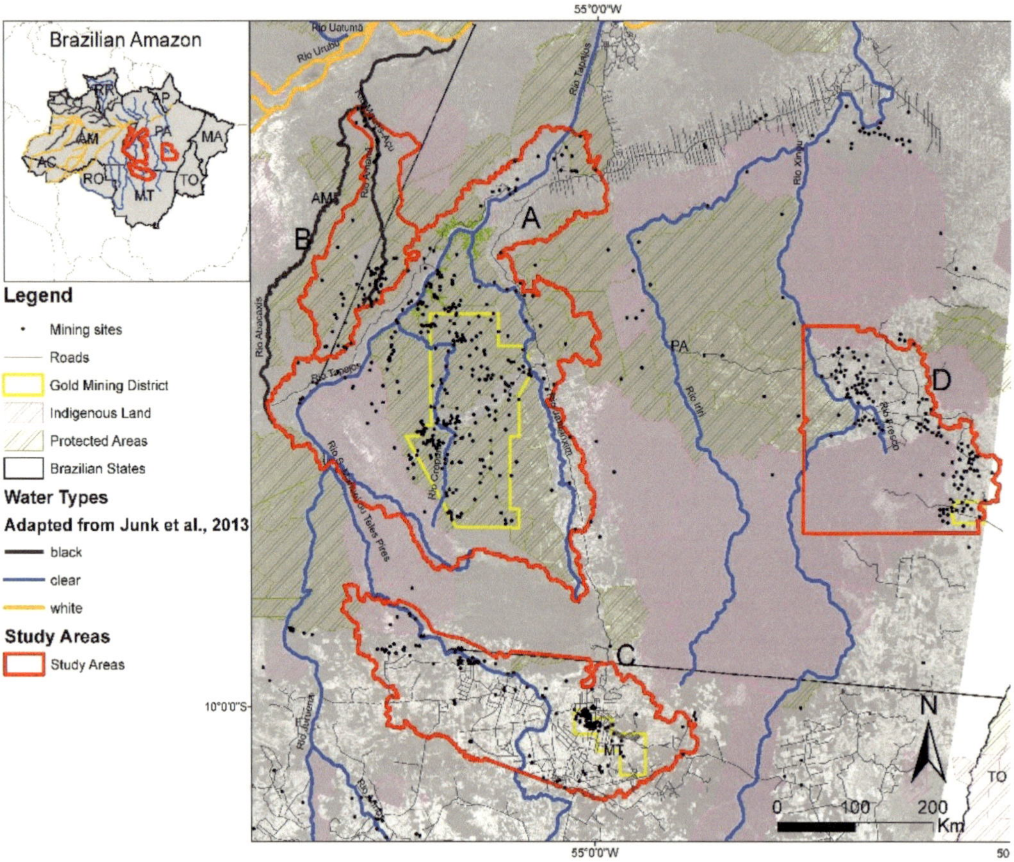

Figure 1. Study areas in the Brazilian Amazon. States of Acre (AC), Amapá (AP), Amazonas (AM), Maranhão (MA), Pará (PA), Rondônia (RO), Roraima (RR), and Tocantins (TO). (A) The Tapajós river basin (PA). (B) The Amanã river (AM). (C) The Teles Pires river, Peixoto Azevedo region (MT). (D) The Fresco river at Xingu river basin (PA).

The second step was to correct the atmospheric effects of original images to surface reflectance. The atmospheric correction (AC) process is necessary for intercalibration of the images in order to compare images from different sensors. The atmospheric correction method for each sensor is chosen according to the output quality and time/processing cost. Only physical-based methods are applied, such as 6S [22], FLAASH [23], ACOLITE [24], and Sen2Cor [25]. For all physical methods, the aerosol optical thickness (AOT) data and water vapor (among other environmental conditions) are required as input to run the model. Some methods retrieve this information from the image (image-based method, such as Sen2Cor and ACOLITE), and others (such as 6S) require the user to indicate this information [26, 27]. The procedure assures that any variation on water-leaving reflectance is due to changes in the water constituents and not to atmospheric effects, neither to intercomparison deviations related to images acquired with different sensors at different times/dates.

The third step was to estimate TSS from corrected satellite images. In this assessment phase, we tested the empirical model developed by Lobo et al. [20] based on in situ radiometric and TSS

http://dx.doi.org/10.5772/intechopen.79725

Satellite/sensor	Life span	Swath (km)	Resolutions				Atm. correction	Available at
			Spatial	Temperature (days)	Radiometric	Spectral		
Landsat-1&2/MSS	1973–1978	170	60 m	16	8 bit	550, 650, 750, and 900 nm	Corrected	earthexplorer.usgs.gov
Landsat-5/thematic mapper (TM)/	1983–1997	170	30 m	16	8 bit	470, 550, 660, 830, 1600, and 2200 nm	Corrected	earthexplorer.usgs.gov
IRS/LISS-III	2011–	141	23 m	5	10 bit	550, 660, 810, and 1600 nm	FLAASH	www.dgi.inpe.br
RapidEye	2012–2015	77	5 m	6	12 bit	470, 550, 660, 705, and 820 nm	FLAASH	geocatalogo.mma.gov.br
Landsat-8/OLI	2014–	170	30 m	16	12 bit	440, 470, 550, 660, 830, 1600, and 2200 nm	ACOLITE or 6S	earthexplorer.usgs.gov
CBERS-4/MUX	2015–	120	20 m	16	8 bit	470, 550, 660, 705, and 820 nm	6S	www.dgi.inpe.br
Sentinel-2/ESA	2015–	290	10 m	5	12 bit	470, 550, 660, 705, and 820 nm	Sen2Cor	earthexplorer.usgs.gov

Table 1. Satellite/sensor specification in terms of resolution, life span, level of atmospheric correction, and source of data.

Satellite/sensor	Red band (ρ_w)		Parameters for Eq. 1		
	Range (nm)	Centered (nm)	*a*	*b*	*c*
Landsat-1&2/MSS	600–700	650	2.272	2.558	2.230
Landsat-5/Thematic Mapper (TM)	630–690	660	2.272	2.468	2.154
Landsat-5/OLI	640–670	655	2.272	2.516	2.182
CBERS-4/MUX	630–690	660	2.272	2.471	2.156
Sentinel-2	640–680	665	2.272	2.469	2.188
IRS/LISS-III	630–690	660	2.272	2.468	2.154
RapidEye	630–685	658	2.272	2.484	2.163

Table 2. Parameters to retrieve TSS from several satellites/sensors using the atmospherically corrected red band.

measurements taken in April 2011 [23 sample points were taken during high water level (water depth ~8 m)] and September 2012 [16 sample points were taken during low water level (water depth ~3 m)]. The measured in situ reflectance (ρw) was resampled for Landsat-5/TM spectral band. Then, to establish an empirical relationship between measured TSS and measured reflectance data, the TSS concentrations measured at 39 sample points were used. To evaluate the use of satellite images on TSS retrieving, this empirical algorithm was applied on two satellite image sets acquired at the same period of field campaigns: Landsat-5/TM (April, 2011) and IRS/LISS-III (September, 2012). To do so, the algorithm was inverted so the user can extract TSS from ρw. In the case of Landsat-5/TM and LISS-III, the following algorithm is applied:

$$TSS = a + (\rho w / b)\, c \tag{1}$$

where ρw is the surface reflectance at red band, $a = 2.272$, $b = 2.468$, and $c = 2.154$.

Considering the different spectral resolution of the orbital sensors, the radiometric measurements were resampled to the sensor's specification during the assessment phase. As a result, specific empirical curves were generated for each sensor (**Table 2**).

After estimating TSS from satellite images, the values were clustered into six TSS concentration classes ranging from 0 to 5, 10, 20, 50, 120, and >120 mgL^{-1}. Finally, the water surface area for each TSS class was tabulated (in km^2) in order to evaluate the amount of water surface with increased TSS values (>20 mgL^{-1}), as opposed to pristine conditions (<20 mgL^{-1}) of clear waters.

4. Estimating TSS concentration from satellite imagery

Results of TSS estimation for study areas A–D (**Figure 1**), from the assessment phase, are presented along with relevant information to characterize SSGM activities and their discharge of sediment into the water.

4.1. The Tapajós river and tributaries

The lower section of the Tapajós river basin (study area A) located in the State of Pará (Brazil) covers about 130,370 km^2. In terms of SSGM, more than 300 small-scale gold mines with more than 50,000 miners cover approximately 230 km^2 [28]. As a result of SSGM activities, the TSS distribution over the Tapajós river and the main tributaries (Crepori and Jamanxim rivers) was extensively presented by Lobo et al. [20]. The authors indicated that the upstream section of the Tapajós river is naturally classified as "clear water" [16, 18, 29]. This class presented relatively low TSS (~5.0 mgL^{-1}), low dissolved organic matter (absorption coefficient for colored dissolved organic matter, a_{cdom} < 2.5 m^{-1}), and low chlorophyll-a concentration (*chl-a* < 1.0 μgL^{-1}), thus resulting in a relatively deep euphotic zone at 1% of total income irradiance ($Z_{1\%}$ ~6.0 m). In these waters, the suspended sediment has a considerable amount of organic matter (~30% of TSS), composed mostly of allochthonous plant debris [30]. The characteristics and concentration of the suspended sediments change abruptly as the Tapajós river receives clay-rich tributaries, such as the heavily mined Crepori river (TSS ~111.3 mgL^{-1}, particulate organic matter <3%, and euphotic depth ~2.0 m). The sediment plume from the Crepori river only fully mixes with the Tapajós river waters at about 200 km downstream after passing through rapids [16, 20]. After these rapids, as the water velocity decreases, the fine suspended solids sink, and concentrations decrease to values similar to those of the upstream Tapajós river (see Telmer et al. [16]). Similar to the Tapajós river, TSS at the Jamanxim river increases as it receives a sediment-rich discharge from the Novo and the Tocantins sub-basins subject to mining operations. In the low water level season (IRS/LISS-III acquired on September 16, 2012), for example, TSS values of about 115.0 mgL^{-1} were estimated for the Crepori river (**Figure 2**).

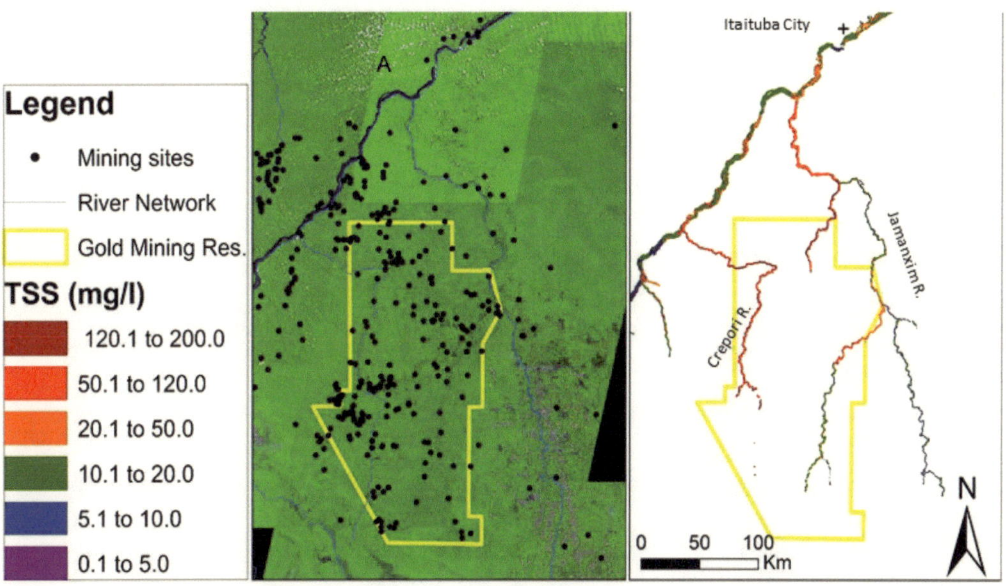

Figure 2. Study area A. TSS concentration with classes ranging from 0.1 to 200 mg L^{-1} using Eq. 1 along the Tapajós river derived from IRS/LISS-III (September 16, 2012).

In terms of surface water impacted by water siltation, **Table 3** shows that Jamanxim river and tributaries present 54% of total surface water (174.3 km^2) with increased TSS levels (>20 mgL^{-1}), mostly because its tributaries Novo and Tocantinzinho rivers present high TSS levels (**Figure 2**). The Crepori river also presents elevated TSS values for 91% of total water surface (31.1 km^2). The discharge of these tributaries into the Tapajós river affects 30% of total surface water (1208.8 km^2) of the Tapajós main channel. These results confirm that Tapajós river and tributaries have been subject to intense water siltation derived from small-scale gold mining that are changing water quality conditions from clear water (<20 mgL^{-1}) in pristine conditions to white waters (TSS levels higher than 20 mgL^{-1}).

Moreover, the recent research applied a sediment modeling approach on Crepori basin to simulate the impacts of past land use-cover change (LUCC) on TSS in the river, comparing these impacts to the effects of gold mining activity in TSS [31]. When comparing the TSS simulated with the 1998–2012 scenario and the estimates conducted by Lobo et al. [20] for the same period, on average, about 14% of TSS estimated by Lobo et al. [20] for high water season is derived by diffuse soil erosion, whereas this proportion is about 6%, on average, for the low water period. Therefore, this suggests that the remaining proportion of TSS measured and estimated by Lobo et al. [20], that is, over 86% of TSS estimated/measured, can be attributed to the gold mining activity.

4.2. The Amanã river

The Amanã river (study area B) is located between the state of Pará and Amazonas, and most of the SSGM is related to Tapajós gold domain as well (Santos et al. [30]). The inclusion of

Study area	River	Extension (km)	Percentage (%) of surface area per class of TSS (mgL^{-1}) from total							Total km^2
			0–5	10	20	50	120	>120	TSS > 20	
A	Tapajos main channel	483.3	27%	43%	27%	3%	0%	0%	30%	1280.8
	Jamanxim and tributaries	707.2	21%	25%	14%	21%	17%	1%	54%	174.3
	Crepori river	228.7	7%	2%	2%	9%	70%	11%	91%	31.1
B	Amana river	95.1	40%	29%	5%	6%	14%	5%	31%	15.1
C	Xingu main channel	185.5	87%	11%	2%	0%	0%	0%	2%	229.6
	Fresco and tributaries	266.7	43%	5%	2%	47%	0%	4%	53%	39.2
D	Teles Pires main channel	120.5	87%	12%	0%	0%	0%	0%	1%	50.2
	Peixoto de Azevedo river	140.4	20%	64%	16%	1%	0%	0%	16%	14.2

Indication of percentage of surface area per class of TSS (mgL^{-1}) from total area mapped.

Table 3. Total water surface mapped for four study areas in the Brazilian Amazon.

http://dx.doi.org/10.5772/intechopen.79725

this area holds on the need for information about SSGM impacts on aquatic systems by the Environmental Protection Agency (ICMBio) for a better management of protected areas in the Amazon region. Recently, a federal police operation shuts down a group of about hundred miners that used to exploit gold illegally for the past 10 years. Local police agents estimate that approximately US$ 10,000 were generated with small-scale gold mining; they also claim that illegal activities caused intense area degradation for more than 70 hectares as well as mercury and cyanide contamination (both used in the gold extraction process) [32].

In terms of water siltation, preliminary results from Landsat-8/OLI (July 29, 2016) indicate TSS values higher than 120 mgL^{-1} where mining sites are present (**Figure 3**). TSS concentration decreases as the river enters the protected area (Flona de Pau-Rosa). In terms of surface water with increased TSS levels, **Table 3** indicates that 31% of 15.1 km^2 presents elevated TSS, mostly located in the upstream section of the river, as shown in **Figure 3**. The TSS values only decrease when the river reaches a more stable and wider section downstream. Overall, these results confirm that the Amanã river has also been subject to intense water siltation derived from small-scale gold mining that are changing water quality conditions from clear water (<20 mgL^{-1}) in pristine conditions to white waters (TSS levels higher than 20 mgL^{-1}).

4.3. Peixoto de Azevedo/Teles Pires river

The Peixoto de Azevedo river (study area C) is located upstream of the Tapajós (so-called Teles Pires river), characterized by natural clear water (TSS < 20 mgL^{-1}). This region is marked also by high deforestation rate, mostly due to the conversion of forest into pasture and agriculture fields [33]. SSGM was intense from 1970 to 1998, but recently the activity is not as intense as it had been before [30]. However, the sediment plume caused either by current mining activity or by degraded areas is still detectable by satellite images (**Figure 4**). The TSS estimation using Landsat-8/OLI (August 06, 2016) for the Peixoto de Azevedo river was up to 20 mgL^{-1} as opposed to the water from the Tapajós upstream (Teles Pires), with TSS lower than 20 mgL^{-1}. The water surface with increased TSS (>20 mgL^{-1}), however, was only 16% for Peixoto de Azevedo (out of 14.2 km^2 mapped) and 1% of Teles Pires river (out of 50.2 km^2). The

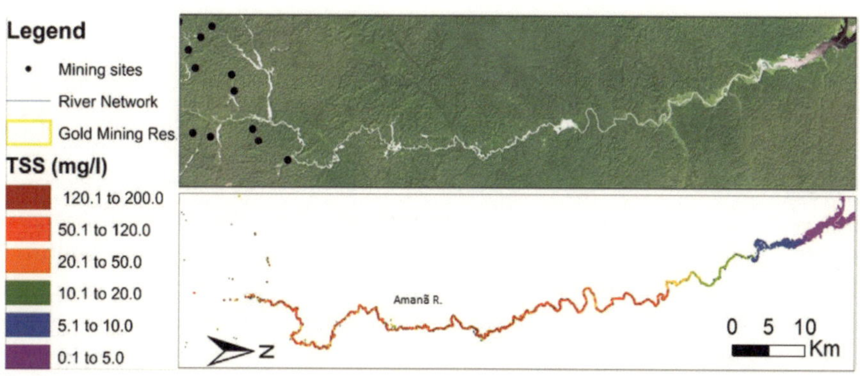

Figure 3. Study Area B (Amanã river). TSS concentration with classes ranging from 0.1 to 200 mgL^{-1} using Eq. 1 along the Amanã river located in the border between Pará and Amazonas states derived from Landsat-8/OLI (July 29, 2016).

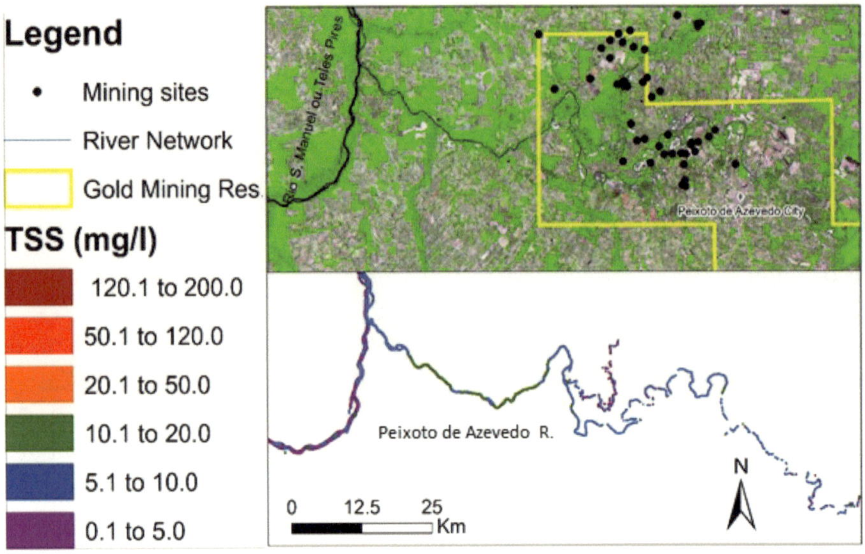

Figure 4. Study area C (Peixoto de Azevedo river/Teles Pires). TSS concentration using Eq. 1 along the Peixoto Azevedo river derived from Landsat-8/OLI (August 06, 2016).

water siltation in the Peixoto de Azevedo river is not too high, indicating that the sediment discharged by current small-scale gold mining is not enough to change the water quality as it occurs in the Tapajós and Amanã rivers.

4.4. Fresco and Xingu river

The Xingu river, likewise the Tapajós river, has its headwaters in the Brazilian central shield, and as a consequence, it also has clear water characterized by low TSS concentration [19]. The Xingu river basin presents several indigenous lands and protected areas that have been threatened by SSGM for decades [34]. Today, intense SSGM is taking place in the Fresco river (**Figure 5**) at the borders of the Kayapó territory. As mining activities within indigenous land are prohibited in Brazil, a recent federal police operation had closed an illegal mining activity in the Kayapó land where drafts, dredges, guns, and mercury were apprehended [35].

As a result of intense SSGM in this region, the TSS estimation using Sentinel-2 image (July 18, 2016) shows TSS higher than 200 mgL^{-1} in the Branco river, which in turn discharges into the Fresco river. At the São Félix do Xingu region, the Fresco river presented TSS concentration of up to 50 mgL^{-1}. For the Xingu Main channel, only 2% of total water surface (229.6 km^2) shows TSS above 10 mgL^{-1}, which corresponds to the sediment-rich Fresco river discharge area. In fact, 39.2 km^2 of the Fresco river and tributaries is analyzed in this study (**Figure 5**); 53% presents TSS above 10 mgL^{-1} (**Table 3**). Once more, the results confirm that Fresco river and tributaries have been subject to intense water siltation derived from small-scale gold mining that are changing water quality conditions from clear water (<20 mgL^{-1}), in pristine conditions, to white waters (TSS levels higher than 20 mgL^{-1}).

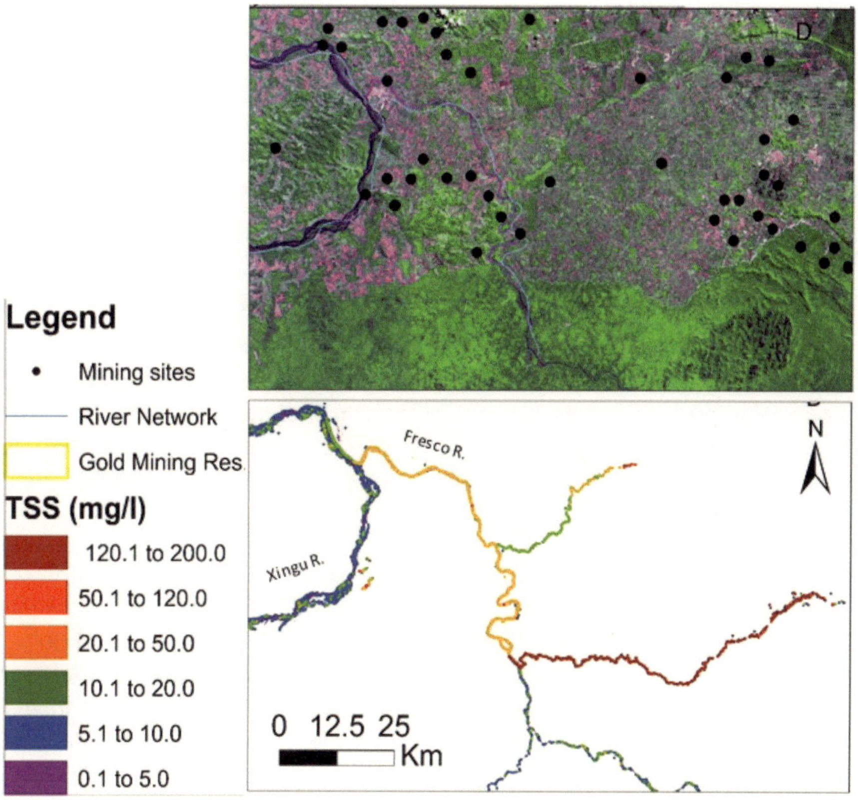

Figure 5. Study area D (Fresco and Xingu rivers). TSS concentration using Eq. 1 along the Fresco river derived from Sentinel-2 (July 18, 2016).

5. Future research

Given the favorable results derived from the first phase of assessment of the methodology, the next steps of this research (second phase) include the establishment of an image-processing framework to correct a large imagery base of multiple sensors in order to build a time series from the 1970s to present, and validation of the empirical model designed by Lobo et al. [20] with field campaigns in the selected areas to estimate TSS from the corrected imagery.

The approach that will be used in the viability assessment for building a time series follows the method applied by Lobo et al. [20], which takes dark dense vegetation (DDV) as reference for atmospheric correction. The main input parameters such as AOT, ozone, and water vapor will be optimized until the forest spectra match those from the imagery calibrated with radiometric measurements (Landsat-5/TM and IRS/LISS-III calibrated with in situ radiometric measurements in April 2011 and September 2012).

Once all images are atmospherically corrected and organized into a database, they are ready for TSS estimation using the parameters in **Table 2**. The challenge here is to validate the application of empirical model designed by Lobo et al. [20] into the additional areas to estimate TSS from the corrected imagery. Although the selected areas present similar characteristics of river basin conditions and mining techniques, slight variation of sediment composition or even the presence of dissolved organic matter can be observed among these areas. Therefore, an effort to sample water in the extended areas (B–D) in order to validate the empirical model is key for a broader application and water quality monitoring purposes. The application of the empirical model to other areas needs a validation process that includes TSS and radiometry acquisition.

Products derived from historical Landsat-1&2/MSS and Landsat-5/TM data (1973–2011) as well as current data from Sentinel-2, Landsat-8/OLI, and RapidEye will be available online at INPE (www.dpi.inpe.br/labisa). One of the main users, and actually who claimed for TSS information within protected areas, is the ICMBio (Chico Mendes Institute for Biodiversity). They have expressed direct interest concerning the SSGM impacts in rivers to support the territory management plan and to indicate areas where urgent action to control illegal activities is needed.

In addition to the direct use for SSGM controlling and monitoring by these national agencies, the applicability of these products for aquatic research is enormous. In fact, evaluation of light attenuation caused by mining tailing has shown that, in general, the euphotic zone in impacted rivers has decreased to at least half of nonimpacted rivers [36]. Several studies can use TSS distribution as an input for hydrological and sediment transportation models; studies on light attenuation and consequences to the biota, as well as socioeconomic studies, will benefit from this research.

6. Conclusions

We hypothesize that because of intensification of mining activities in the Brazilian Amazon, clear water rivers such as Tapajós and Xingu rivers and its tributaries are becoming or may become permanently turbid waters (so-called white waters in the Amazonian context). This chapter informs the main activities carried out to develop a monitoring system for quantifying water siltation caused by SSGM in the Amazon rivers using multi-satellite data in order to investigate this hypothesis.

As a result of the first assessment phase, a multi-satellite approach was developed based on TSS algorithm proposed by Lobo et al. [20]. To do so, radiometric in situ data were resampled to several sensors' specifications, such as Landsat-8/OLI and Sentinel-2, and applied to clear water rivers subject to intense sediment discharged by small-scale gold mining in order to recover TSS concentration. Except for Peixoto de Azevedo (study area C), the results confirm that Tapajós (A), Amanã (B), and Fresco river and tributaries (D) have been subject to intense water siltation derived from small-scale gold mining that are changing water quality conditions from clear water (<20 mgL^{-1}), in pristine conditions, to white waters (TSS levels higher than 20 mgL^{-1}).

In order to establish a monitoring system of water siltation, the next steps of this research (second phase) include an image-processing framework to correct a large imagery base of multiple sensors and validation of the empirical model designed by Lobo et al. [20] with field campaigns in the selected areas to estimate TSS from the corrected imagery.

Acknowledgements

The authors acknowledge financial support from FAPESP (Process Nos. 2011/23594-8 and 2008/07537-1) and CNPq (Process Nos. 237930/2012-9 and 150835/2015-9). We would like to thank all the LabISA (Instrumentation Laboratory for Aquatic Systems at INPE) staff, particularly to Lino de Carvalho and Felipe Menino Carlos, for their support on data processing. Thanks to Lauren Pansegrouw for the English revision.

Conflict of interest

The authors declare no conflict of interest.

Author details

Felipe de Lucia Lobo[1]*, Evlyn Márcia Leão de Moraes Novo[1],
Claudio Clemente Faria Barbosa[2] and Vitor Hugo Fernandes de Vasconcelos[3]

*Address all correspondence to: felipe.lobo@inpe.br

1 Remote Sensing Division, National Institute for Space Research (INPE), São José dos Campos, SP, Brazil

2 Image Processing Division, National Institute for Space Research (INPE), São José dos Campos, SP, Brazil

3 Monitoring and Environmental Information Division (DMIF/CGPRO), Instituto Chico Mendes (ICMBio), Brasília/DF, Brazil

References

[1] Goulding M, Barthem R, Ferreira EJ. Smithsonian Atlas of the Amazon. Washington, D.C.: Smithsonian; 2003. ISBN: 9781588341358

[2] Fearnside PM. Dams in the Amazon: Belo monte and Brazil's hydroelectric development of the Xingu River Basin. Environmental Management. 2006;**38**:16-27. DOI: 10.1007/s00267-005-0113-6

[3] Fearnside PM. Environmental Destruction in the Brazilian Amazon. The Future of Amazonia: Destruction or Sustainable Development; 1990. pp. 179-225

[4] Fearnside PM. A hidrelétrica de São Luiz do Tapajós: A arte de EIAS cosméticos. Hidrelétricas Na Amazônia. 2015;**8**:115

[5] Finer M, Novoa S. Gold Mining Deforestation Continues to Expand in la Pampa (Madre De Dios, Peru). Available from: http://maaproject.org/2015/03/gold-mining-deforestation-expand-peruvian-amazon/ [Accessed: January 26, 2016]

[6] Murguía DI, Bringezu S, Schaldach R. Global direct pressures on biodiversity by large-scale metal mining: Spatial distribution and implications for conservation. Journal of Environmental Management. 2016;**180**:409-420. DOI: 10.1016/j.jenvman.2016.05.040

[7] Veiga MM, Hinton JJ. Abandoned artisanal gold mines in the Brazilian Amazon: A legacy of mercury pollution. Natural Resources Forum. 2002;**26**:15-26. DOI: 10.1111/1477-8947.00003

[8] Coelho MCN, Wanderley LJ, Costa RC. Small scale gold mining in the XXI century. Examples in the south-west Brazilian Amazon. Anuário do Instituto de Geociências—UFRJ. 2016;**39**:10

[9] Neto A, do Ouro HP. Ministério de Minas e Energia. Brasília; 2009. p. 50

[10] Silva AR. A indústria mineral no pará. Belém. 2012:178

[11] Fernandes FRC, Alamino RDCJ, Araújo ER. Recursos minerais e comunidade: Impactos humanos, socioambientais e econômicos. Vol. 1. Rio de Janeiro: CETEM/MCTI; 2014

[12] Rodrigues RM, Mascarenhas AFS, Ichihara AH, Souza TMC. Estudo dos impactos ambientais decorrentes do extrativismo mineral e poluição mercurial no tapajós - pré-diagnóstico. Vol. Série Tecnologia Ambiental. Rio de Janeiro: CETEM/CNPq; 1994. p. 220

[13] Roland F, Esteves FD. Effects of bauxite tailing on PAR attenuation in an amazonian crystalline water lake. Hydrobiologia. 1998;**377**:1-7. ISSN: 0018-8158

[14] Tudesque L, Grenouillet G, Gevrey M, Khazraie K, Brosse S. Influence of small-scale gold mining on French Guiana streams: Are diatom assemblages valid disturbance sensors? Ecological Indicators. 2012;**14**:100-106. DOI: 10.1016/j.ecolind.2011.07.018

[15] Mol JH, Ouboter PE. Downstream effects of erosion from small-scale gold mining on the instream habitat and fish community of a small neotropical rainforest stream. Conservation Biology. 2004;**18**:201-214. ISSN: 0888-8892

[16] Telmer K, Costa M, Simões Angélica R, Araujo ES, Maurice Y. The source and fate of sediment and mercury in the Tapajós River, Pará, Brazilian Amazon: Ground- and space-based evidence. Journal of Environmental Management. 2006;**81**:101-113

[17] Telmer K, Veiga M. World Emissions of Mercury from Artisanal and Small Scale Gold Mining; 2009

[18] Sioli H. The Amazon and its main affluents: Hydrography, morphology of the river courses, and river types. In: Sioli H, editor. The Amazon. Vol. 56. Netherlands: Springer; 1984. pp. 127-165. DOI: 10.1007/978-94-009-6542-3_5

[19] Junk WJ, Piedade MTF, Schöngart J, Cohn-Haft M, Adeney JM, Wittmann F. A classification of major naturally-occurring amazonian lowland wetlands. Wetlands. 2011;**31**:623-640. DOI: 10.1007/s13157-011-0190-7

[20] Lobo FL, Costa MPF, Novo EM. Time-series analysis of LANDSAT-MSS/TM/OLI images over amazonian waters impacted by gold mining activities. Remote Sensing of Environment. 2015;**157**:170-184. DOI: 10.1016/j.rse.2014.04.030

[21] SiBCS. Sistema brasileiro de classificação de solos. 2nd ed. Rio de Janeiro: Embrapa Solos; 2006

[22] Vermote EF, Tanre D, Deuze JL, Herman M, Morcette JJ. Second simulation of the satellite signal in the solar spectrum, 6s: An overview. IEEE Transactions on Geoscience and Remote Sensing. 1997;**35**:675-686. DOI: 10.1109/36.581987

[23] Envi. Envi FLAASH version. 4.1, September, 2004 ed. Research Systems, Inc; 2004. https://www.harrisgeospatial.com/docs/BackgroundFLAASH.html [Accessed: June 19, 2018]

[24] Vanhellemont Q, Ruddick K. Turbid wakes associated with offshore wind turbines observed with Landsat 8. Remote Sensing of Environment. 2014;**145**:105-115. DOI: 10.1016/j.rse.2014.01.009

[25] ESA. S2 Mpc Level 2a Product Format Specification. European Space Agency; 2016. p. 39

[26] Pahlevan N, Schott JR. Leveraging eo-1 to evaluate capability of new generation of landsat sensors for coastal/inland water studies. IEEE Journal of Selected Topics in Applied Earth Observations and Remote Sensing. 2013;**6**:360-374. DOI: 10.1109/jstars.2012.2235174

[27] Martins V, Barbosa C, de Carvalho L, Jorge D, Lobo F, Novo E. Assessment of atmospheric correction methods for Sentinel-2 MSI images applied to amazon floodplain lakes. Remote Sensing. 2017;**9**:322. DOI: 10.3390/rs9040322

[28] Lobo F, Costa M, Novo E, Telmer K. Distribution of artisanal and small-scale gold mining in the Tapajós River Basin (Brazilian Amazon) over the past 40 years and relationship with water siltation. Remote Sensing. 2016;**8**(7):22. DOI: 10.3390/rs8070579

[29] Junk W. General aspects of floodplain ecology with special reference to amazonian floodplains. In: Junk W, editor. The Central Amazon Floodplain. Vol. 126. Berlin Heidelberg: Springer; 1997. pp. 3-20. DOI: 10.1007/978-3-662-03416-3_1

[30] Santos JOS, Groves D, Hartmann L. Gold deposits of the Tapajós and Alta Floresta domains, Tapajós–Parima orogenic belt, Amazon Craton, Brazil. Mineralium Deposita. 2001;**36**:278-299. DOI: 10.1007/s001260100172

[31] Abe C, Novo E, Lobo F, Costa MPF, Dibike Y. Impact of land use change on the hydrology and sediment load in an amazonian basin using SWAT. in prep., Regional Environmental Change Journal, currently under revision. 26

[32] Melo K. Exploradores movimentaram R$ 30 milhões com mineração irregular no amazonas. Available from: http://amazonia.org.br/2015/09/exploradores-movimentaram-r-30-milhoes-com-mineracao-irregular-no-amazonas/ [Accessed: June 12, 2018]

[33] Almeida CAd, Coutinho AC, Esquerdo JCDM, Adami M, Venturieri A, Diniz CG, Dessay N, Durieux L, Gomes AR. High spatial resolution land use and land cover mapping of the Brazilian legal Amazon in 2008 using Landsat-5/TM and MODIS data. Acta Amazonica. 2016;**46**:291-302. ISSN: 0044-5967

[34] Ricardo F, Rolla A. Mineração em terras indígenas na Amazônia Brasileira. Vol. 1. São Paulo: Instituto Socioambiental; 2005. p. 181

[35] IBAMA. Ibama desativa 29 balsas e máquinas de garimpo na terra indígena Kayapó (PA). Available from: http://www.ibama.gov.br/noticias/422-2017/1230-ibama-desativa-29-balsas-de-garimpo-e-maquinas-escavadeiras-na-terra-indigena-kayapo-pa [Accessed: June 12, 2018]

[36] Lobo F, Costa M, Novo E, Telmer K. Effects of small-scale gold mining tailings on the underwater light field in the tapajós river basin, Brazilian Amazon. Remote Sensing. 2017;**9**:861. DOI: 10.3390/rs9080861

Chapter 6

Limnological Patterns in a Large Subtropical Reservoir Cascade

Marcos Gomes Nogueira and Juliana Pomari

Additional information is available at the end of the chapter

http://dx.doi.org/10.5772/intechopen.80632

Abstract

The study identified limnological patterns in the Paranapanema River reservoir cascade, one of the main tributaries of the high Paraná River, La Plata Basin, southeast Brazil. Samplings were carried out in eight reservoirs from a total of 37 sites. We analyzed the water transparency, depth, and vertical profiles of temperature, pH, conductivity, dissolved oxygen, biochemical oxygen demand, total solids, suspended solids, total nitrogen, total phosphorus, chlorophyll *a*, and thermotolerant coliforms. Additionally, the trophic state index for tropical/subtropical reservoirs, the water retention time and morphometric characteristics of each reservoir were calculated. Longitudinal compartmentalization is conspicuous in storage reservoirs, whereas the magnitude of temporal changes is higher in run-of-river systems. The lateral component of spatial heterogeneity was also very important for some reservoirs, determined basically by the entrance of tributary rivers. On the vertical dimension, summer thermal stratification, followed by oxygen decrease in bottom layers, in the central channel and lacustrine zones of deeper and larger reservoirs was observed. The ultraoligotrophic condition prevailed, despite signals of intensive land use for agriculture—recurrent high phosphorus values. The acquired experience provided a baseline for a permanent limnological and water quality program, which subsidizes management actions in the basin.

Keywords: water quality, trophic state, spatial variability, seasonal variability, run-of-river, storage

1. Introduction

River damming for hydropower production is presently a major human interference on fluvial systems, all over the world. In these first decades of the twenty-first century, the construction of 3700 dams is expected, increasing the global hydropower production by 73%. This would correspond to an intensification in the exploitation of the technically feasible hydropower potential from 22 to 39% [1].

Hydropower is crucial in Brazil, representing the main source of electric generation, 107,601 MW of a total of 166.872 MW, according to the database of the Brazilian National Agency of Electric Energy (http://www.aneel.gov.br). The production is distributed in 1321 reservoirs, 220 large ones (>30 MW) and 1101 smaller ones (<30 MW).

Hydroelectric dams are an integrated and complex generation system that provides relatively clean and renewable energy. However, independent of the reservoir design and operation, they promote substantial interference on the ecological structure and functioning of river basins [2] and permanent loss of the regional geodiversity [3, 4].

Classical studies on reservoirs [5, 6] distinguish at least three limnological zones in large reservoirs: the region with strong influence of the incoming river—lotic region; an intermediate transition region; and finally a region similar to lakes, near the dam. However, it is known that in large tropical reservoirs, this multidimensional zonation is greatly influenced by the input of secondary tributaries and by the differential water retention time of each reservoir arm [7–10]. These distinctive areas or compartments in reservoir ecosystems may also change seasonally, in a complex dynamic [5, 7, 11, 12].

The large-scale variability of reservoirs along the main axis is determined by longitudinal gradients of flow velocity, depth, width, particle sedimentation, transparency and light penetration, and thermal stratification [11–13]. Compared to lakes, reservoirs exhibit high watershed area/water body area, shorter but varying retentions times, a rapid aging process related to watershed uses, and high capability to retain organic and inorganic matter [14, 15].

An integrated approach, including continuous monitoring and innovative researches, is required for the understanding of the reservoir ecosystems, giving their inherent complexity, with many components and subsystems that interact intensively in space and time [16].

The present study was planned to discriminate the main limnological patterns of a tropical/subtropical reservoir cascade aiming to provide scientific support for management purposes. We analyzed the morphometric characteristics; operation type; water retention time; physical, chemical, and biological variables as well as the trophic status. The acquired experience served as the base for the establishing of methodological protocols that have been employed in the long-term (permanent) limnological and water quality monitoring program (8 years now) in the Paranapanema River reservoir cascade (southeast Brazil). The two most important premises are: (1) the reservoir operation (run-of-river or storage systems) is determinant for the establishing of distinct limnological patterns and (2) the Paranapanema River reservoir cascade is influenced by the intensification of the agriculture in the watershed.

2. Study area

The Paranapanema River is one of the main tributaries of the Paraná River (La Plata Basin), located between the coordinates 22–26° S and 47–54° W, on the tropical/subtropical boundary. The river, with a length of 929 km, is under federal jurisdiction, because it is the natural border between the states of São Paulo and Paraná.

Since the 1950s, 11 hydropower plants were constructed in the main river course. In last two decades, these reservoirs (eight larger ones) have been intensively studied by researchers of the São Paulo State University, Campus of Botucatu, São Paulo [17, 18].

The reservoirs selected for the study are Jurumirim (JR), Chavantes (CH), Salto Grande (SG), Canoas II (CII), Canoas I (CI), Capivara (CP), Taquaruçu (TQ), and Rosana (RS), arranged in a cascade (upstream → downstream) system. Three of the them, Jurumirim, Chavantes, and Capivara, are storage systems (i.e., with high water retention times), whereas the others are run-of-river systems. The total installed potential is 2241 MW.

For the development of the study, information was obtained at 37 sampling sites (**Figure 1**), whose distribution intended to cover the entire river's continuous (inter-reservoir) variability as well as the internal (intra-reservoir) longitudinal gradient established between the lotic (Paranapanema River entrance) and lentic (dam) areas of each reservoir. Additionally, the influence of important secondary tributaries (river mouths) was also considered. At

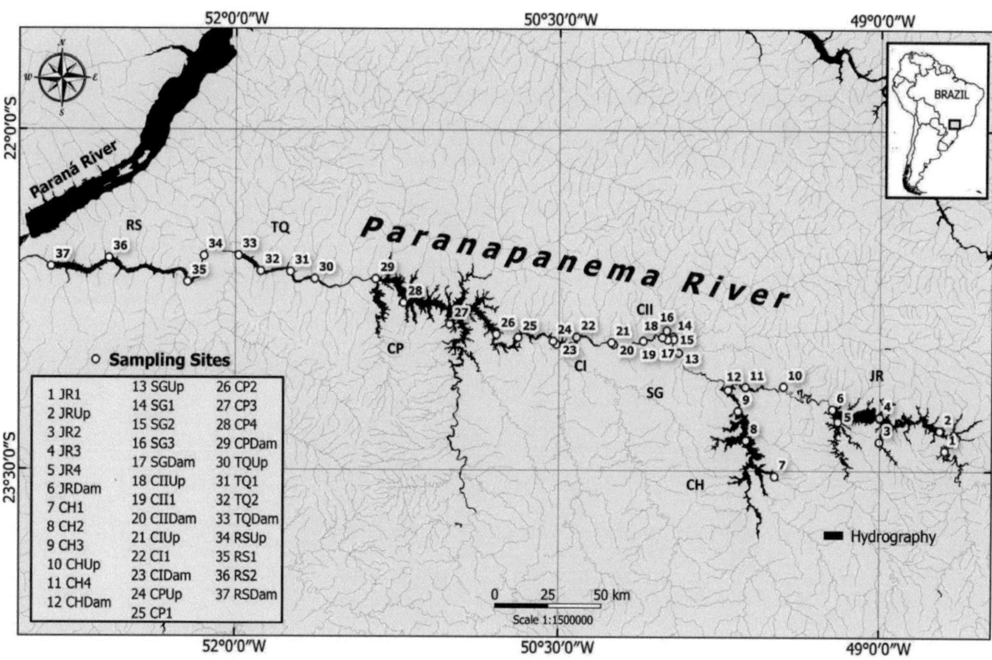

Figure 1. Geographic location of Paranapanema River reservoir cascade and the selected sampling sites.

Reservoir	Sampling site	Coordinates		Altitude (m) (a.s.l.)
Jurumirim	JR1	23°24'39.40"S	48°41'53.80"W	565
	JRUp	23°19'21.90"S	48°43'19.00"W	568
	JR2	23°22'20.80"S	49°0'7.80"W	567
	JR3	23°15'55.40"S	49°0'1.10"W	560
	JR4	23°17'1.90"S	49°11'56.80"W	570
	JRDam	23°13'44.80"S	49°13'29.20"W	563
Chavantes	CH1	23°31'29.50"S	49°29'29.30"W	469
	CH2	23°21'58.90"S	49°37'37.90"W	471
	CH3	23°14'15.00"S	49°39'48.60"W	467
	CHUp	23°7'46.20"S	49°27'2.30"W	468
	CH4	23°7'55.10"S	49°37'47.30"W	477
	CHDam	23°8'41.70"S	49°42'32.40"W	473
Salto Grande	SGUp	22°58'52.60"S	49°56'22.10"W	383
	SG1	22°54'43.60"S	49°57'57.20"W	383
	SG2	22°54'49.60"S	49°58'0.30"W	389
	SG3	22°53'3.20"S	49°59'40.40"W	385
	SGDam	22°53'56.00"S	49°59'27.00"W	389
Canoas II	CIIUp	22°55'5.50"S	49°59'31.50"W	368
	CII1	22°55'42.50"S	50°6'24.00"W	364
	CIIDam	22°56'36.80"S	50°14'42.80"W	367
Canoas I	CIUp	22°56'8.60"S	50°15'19.70"W	348
	CI1	22°54'44.40"S	50°25'5.40"W	348
	CIDam	22°56'35.90"S	50°30'43.10"W	354
Capivara	CPUp	22°55'49.30"S	50°31'38.90"W	329
	CP1	22°54'52.60"S	50°41'22.10"W	326
	CP2	22°54'4.50"S	50°47'24.50"W	322
	CP3	22°51'26.80"S	51°0'22.00"W	335
	CP4	22°45'45.80"S	51°13'10.40"W	332
	CPDam	22°39'27.90"S	51°20'49.80"W	325
Taquaruçu	TQUp	22°39'27.00"S	51°37'46.00"W	277
	TQ1	22°37'36.20"S	51°44'27.00"W	273
	TQ2	22°37'28.00"S	51°52'46.60"W	273
	TQDam	22°33'19.80"S	51°59'7.90"W	271
Rosana	RSUp	22°33'27.50"S	52° 8'59.90"W	260
	RS1	22°40'20.30"S	52°13'32.80"W	268

Reservoir	Sampling site	Coordinates		Altitude (m) (a.s.l.)
	RS2	22°34'1.10"S	52°35'38.60"W	251
	RSDam	22°36'14.30"S	52°51'42.10"W	241

Table 1. Geographic coordinates and elevations of the sampling sites along the reservoir cascade.

least three to six sampling sites were selected in each reservoir. **Table 1** presents the geographic coordinates as well as the altitude data of the sampling sites, obtained with a Garmin Etrex Vista H GPS.

3. Material and methods

The sampling campaigns for physical, chemical, and biological measurements were carried out in two periods of the year: March—wet season (late summer)—and October—dry season (spring) of 2011. All sites of each reservoir/period were sampled in the same or in two consecutive days.

The limnological variables measured *in situ*, and respective methodologies, are shown in **Table 2**.

Water samples were collected with a Van Dorn bottle at three depths along the water column, corresponding to the surface, middle, and bottom (about 0.5–1.0 m above sediments), for the analyses of nutrient (total nitrogen and phosphorus), total solids, suspended solids, chlorophyll *a*, biochemical oxygen, and thermotolerant coliforms. When filtration was required, Millipore AP40 membranes and vacuum pump were used. For weight determination, a Denver analytical scale (0.00001 g) was used. Laboratory determinations followed methodological principles presented in **Table 3**.

The shore line development index was calculated according to [26]. The theoretical residence time (days) was defined as the ratio of reservoir volume and the flow calculated using the formula: $TRT = V/(Q \times 86{,}400)$, where V = reservoir volume (m^3); Q = mean flow ($m^3\ s^{-1}$); and 86,400 = number of seconds contained in a day. Input data for each reservoir are available at the National Water Agency (Agência Nacional de Águas, in Portuguese) (http://sar.ana.gov.br).

Variable	Methodology
Water transparency (Transp)	Secchi disk depth
Water temperature (Temp)	Eureka multi-parameter probe—vertical profile
Dissolved oxygen (DO)	Eureka multi-parameter probe—vertical profile
pH	Eureka multi-parameter probe—vertical profile
Conductivity (Cond)	Eureka multi-parameter probe—vertical profile
Turbidity (Turb)	Eureka multi-parameter probe—vertical profile

Table 2. Limnological variables measured *in situ*.

The trophic state index was determined in accordance with [27] for tropical/subtropical reservoirs (TSItsr), which includes six categories: (U) ultraoligotrophic (≤51.1), (O) oligotrophic (51.2–53.1), (M) mesotrophic (53.2–55.7), (E) eutrophic (55.8–58.1), (S) supereutrophic (58.2–59), and (H) hypereutrophic (≥59.1), respectively.

The studied parameters (when applicable) were compared with the standard references (**Table 4**) established by the federal framework directive for Class 1 waters—the best possible condition (after the Special Class), with no human use restriction and appropriated for communities' protection (http://www.mma.gov.br/port/conama/res/res05/res35705).

A principal component analysis (PCA) was performed to summarize variation tendencies or patterns for limnological variables during both sampling periods, using the PRIMER v6 statistics package for Windows (Plymouth Routines in Multivariate Ecological Research, www.primer-e.com).

Variable	**Methodology**
Total nitrogen (TN)	Spectrophotometry [19, 20]
Total phosphorus (TP)	Spectrophotometry [19, 21]
Total solids (TS)	Evaporation/gravimetry (100°C) [22]
Suspended solids (SS)	Gravimetry [23]
Biochemical oxygen demand (BOD)	Incubation 20°C/5 days [22]
Chlorophyll *a* (Chl)	Spectrophotometry [24, 25]
Thermotolerant coliforms (Coli)	Multiple tube MPN [22]

Table 3. Limnological variables measured in the water samples—laboratory analyses.

Variable	**Standard reference**
pH	6–9
Turbidity	>40 NTU
Dissolved oxygen	>6 mg L^{-1} O_2
Biochemical oxygen demand	>3 mg L^{-1} O_2
Total solids	>500 mg L^{-1}
Total nitrogen	>1.27 mg L^{-1} lentic systems
	>2.18 mg L^{-1} lotic systems
Total phosphorus	>0.020 mg L^{-1} lentic systems
	>0.025 mg L^{-1} intermediate systems*
Chlorophyll *a*	>10 µg L^{-1}
Thermo tolerant coliforms	>200 NMP 100 mL^{-1}

*Water retention time between 2 and 40 days.

Table 4. Standard references established by CONAMA Resolution 357/2005 (conditions for water quality categories for Brazilian aquatic ecosystems) for Class 1 waters.

4. Results[1]

The general characteristics of the studied reservoirs (morphometry, operation, trophy, etc.) are compiled in **Table 5** and the wide spectrum of conditions certainly influence not only the structure but also the limnological functioning of these ecosystems, as pointed out earlier.

The obtained results, *in situ* and laboratory measurements, are presented following the reservoir's sequence, from upstream toward the mouth. Graphical representations (**Figures 2–33**) are standardized for all reservoirs, including monthly variation of the water level versus flow, vertical profiles of temperature and dissolved oxygen, and intra-reservoir variability (upstream/tail → dam) of transparency, electric conductivity, pH, turbidity, total solids, suspended solids, biochemical oxygen demand, total nitrogen, total phosphorus, chlorophyll *a*, and thermotolerant coliforms.

Variable	JR	CH	SG	CII	CI	CP	TQ	RS
Type (*modus operandi*)*	S	S	R	R	R	S	R	R
Area (km^2)	449	400	12	22.5	30.85	576	80.1	220
Perimeter (km)	1286	1085	81	103	120	1550	301	433
Volume (hm^3)	7702	9410	63.2	158	220	11,743	754.2	1942
Shore line development index	17.1	15.3	6.6	6.1	6.1	18.2	9.5	8.2
Water retention time (days)	400	335	2	4	6	115	7	17
Maximum measured depth (m)	32	79	10	15	25	40	26	27
Altitude (m) (a.s.l.)	563	473	389	367	354	325	271	241
Age (year)	56	47	60	19	19	40	29	31
T.S.I. $_{tsr}$**	U	U	U	U	O	U	U	U

*Reservoir operation S—Storage; R—Run-of-River.
**Trophic State Index $_{tropical/subtropical\ reservoir}$ (U = ultraoligotrophic; O = oligotrophic).
Minimum (dark gray) and maximum (light gray) values are highlighted.

Table 5. General characteristics of the study reservoirs.

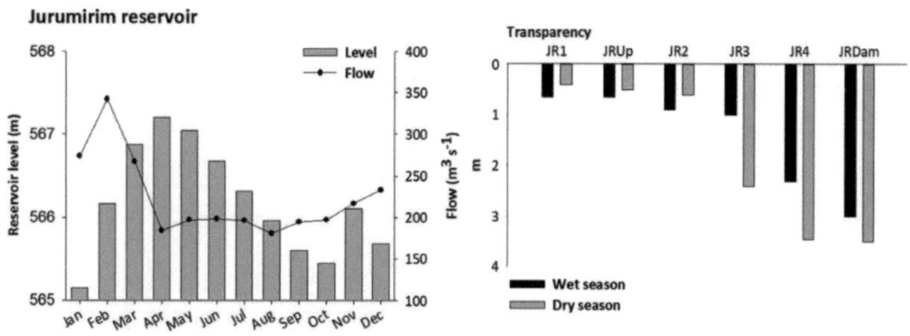

Figure 2. Variation of water level and flow (monthly averages) and transparency (wet and dry seasons) in Jurumirim reservoir.

[1]See also Appendixes 1 and 2.

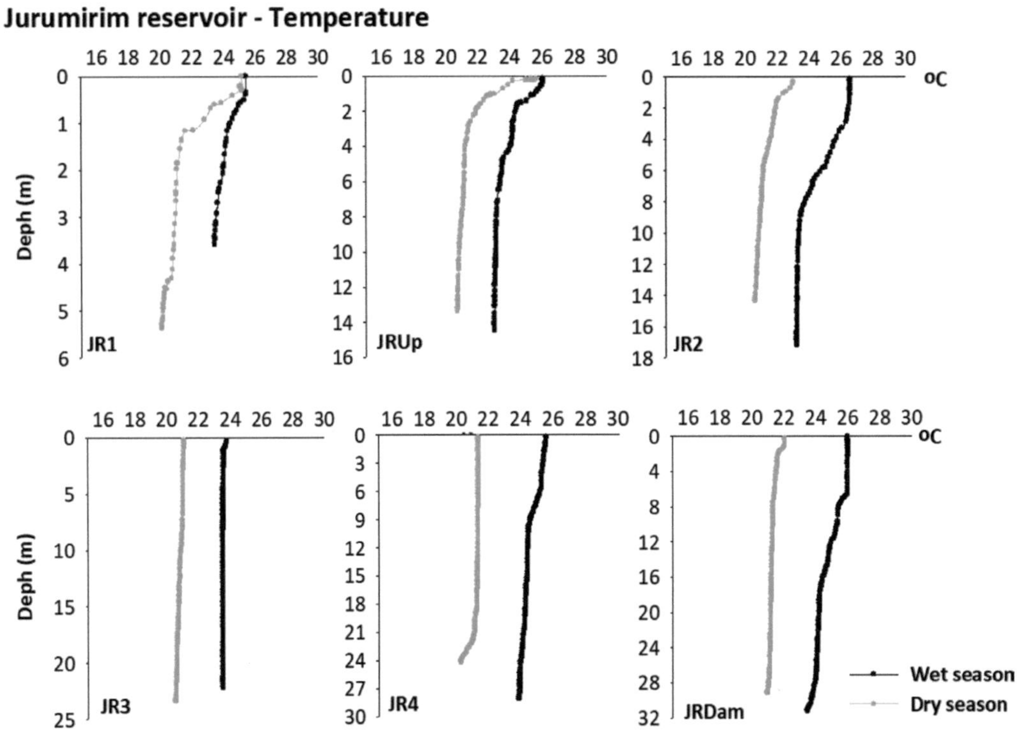

Figure 3. Temperature profiles in Jurumirim reservoir sampling sites during wet and dry seasons.

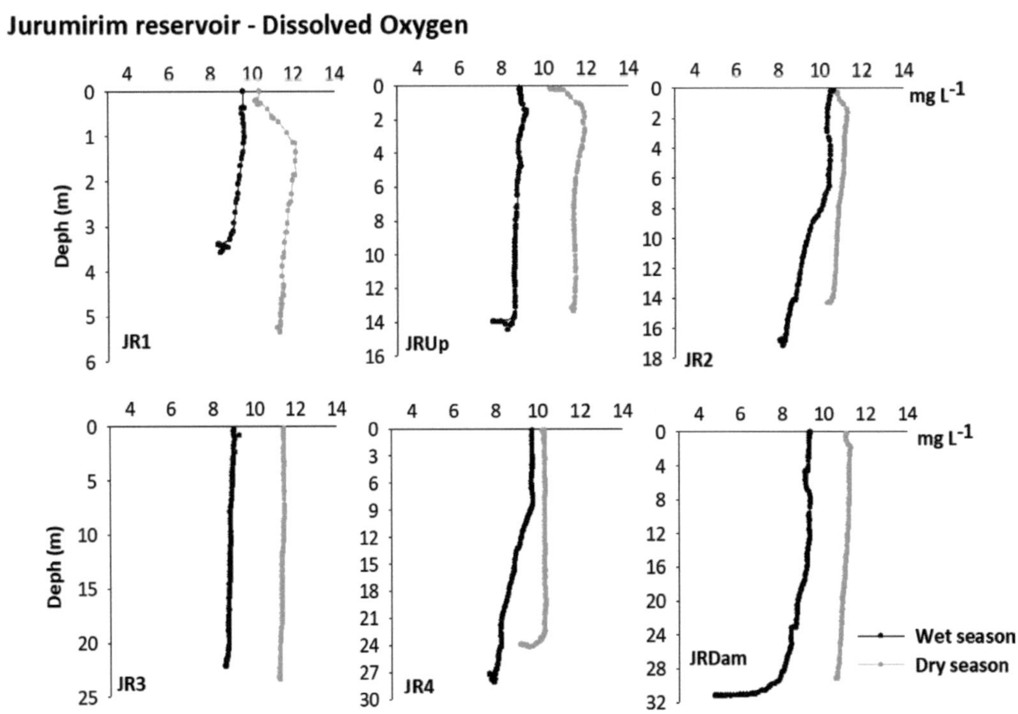

Figure 4. Dissolved oxygen profiles in Jurumirim reservoir sampling sites during wet and dry seasons.

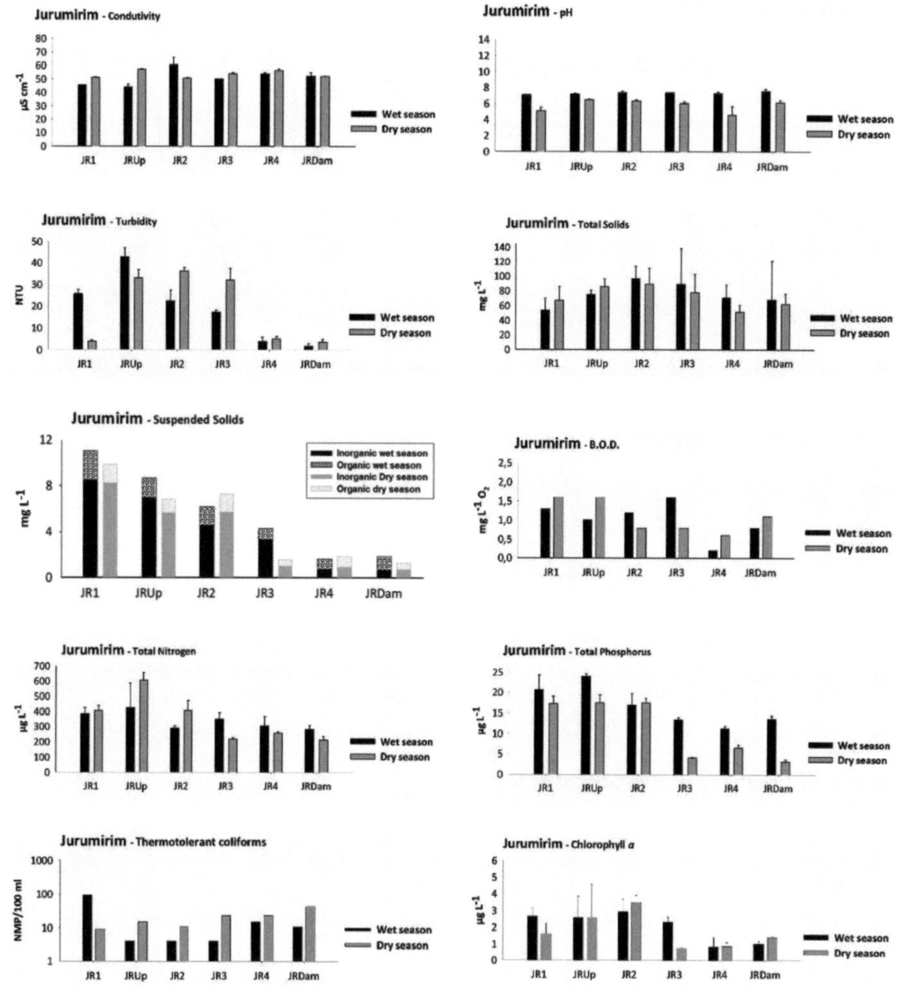

Figure 5. Mean values (+S.D.) of conductivity, pH, turbidity, total and suspended solids, biochemical oxygen demand, total nitrogen, total phosphorus, thermo tolerant coliforms and chlorophyll *a* in Jurumirim reservoir sampling sites during wet and dry seasons.

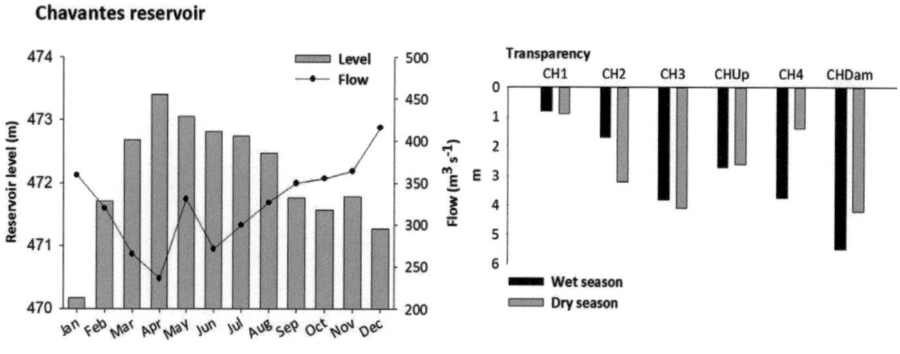

Figure 6. Variation of water level and flow (monthly averages) and transparency (wet and dry seasons) in Chavantes reservoir.

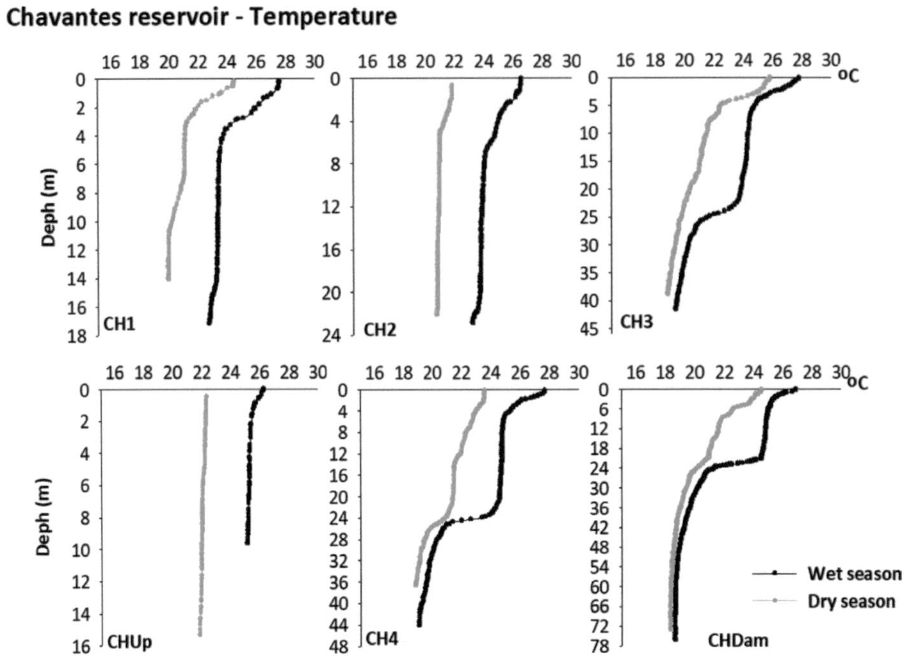

Figure 7. Temperature profiles in Chavantes reservoir sampling sites during wet and dry seasons.

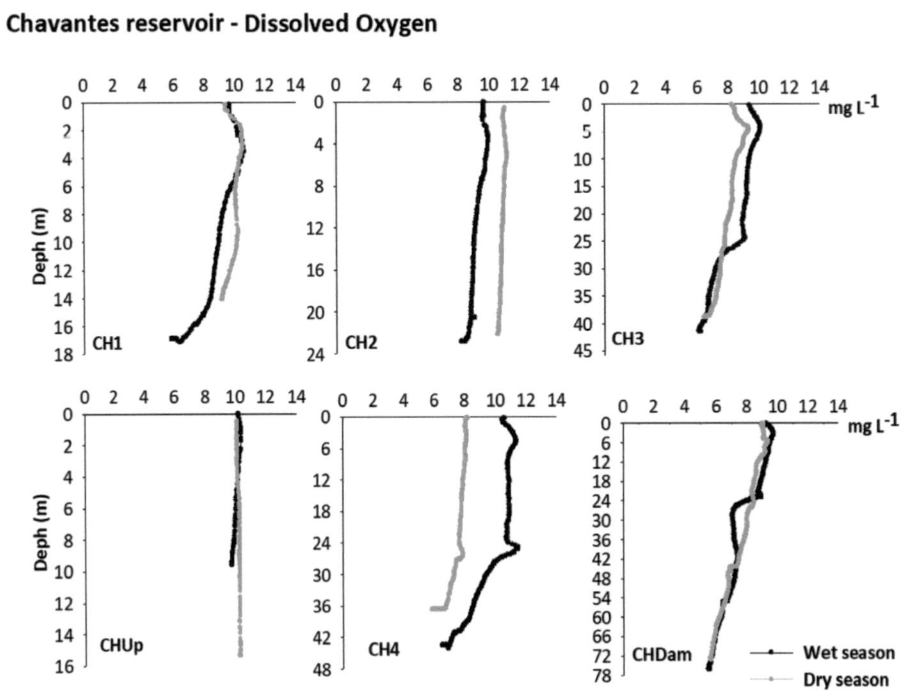

Figure 8. Dissolved oxygen profiles in Chavantes reservoir sampling sites during wet and dry seasons.

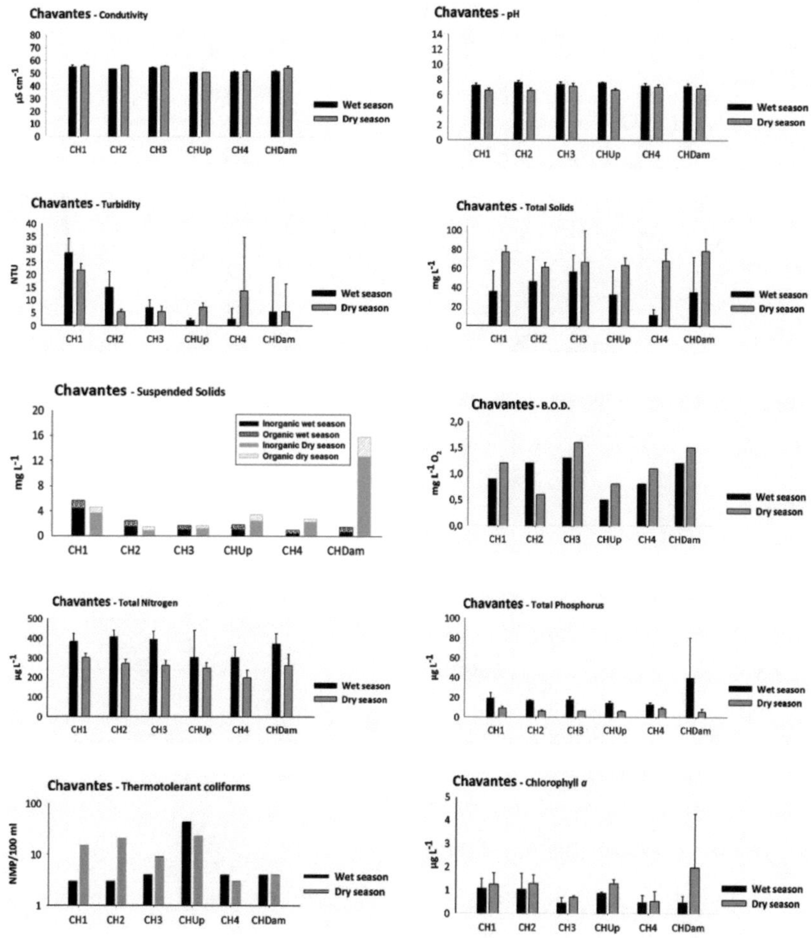

Figure 9. Mean values (+S.D.) of conductivity, pH, turbidity, total and suspended solids, biochemical oxygen demand, total nitrogen, total phosphorus, thermotolerant coliforms and chlorophyll *a* in Chavantes reservoir sampling sites during wet and dry seasons.

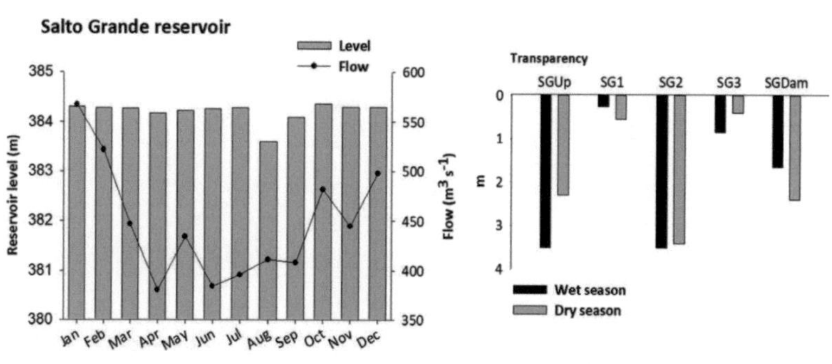

Figure 10. Variation of water level and flow (monthly averages) and transparency (wet and dry seasons) in Salto Grande reservoir.

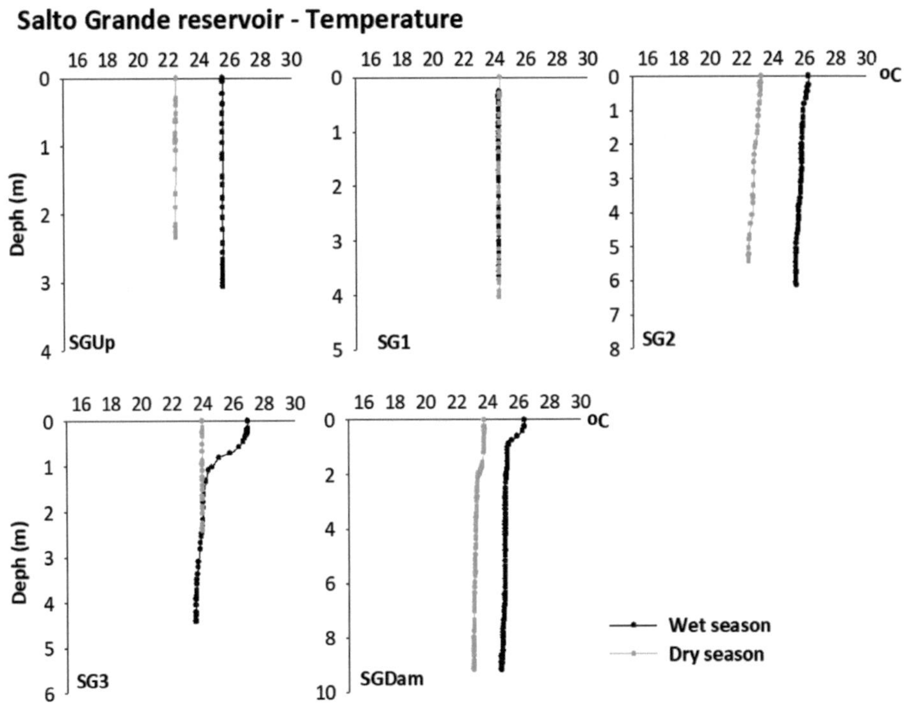

Figure 11. Temperature profiles in Salto Grande reservoir sampling sites during wet and dry seasons.

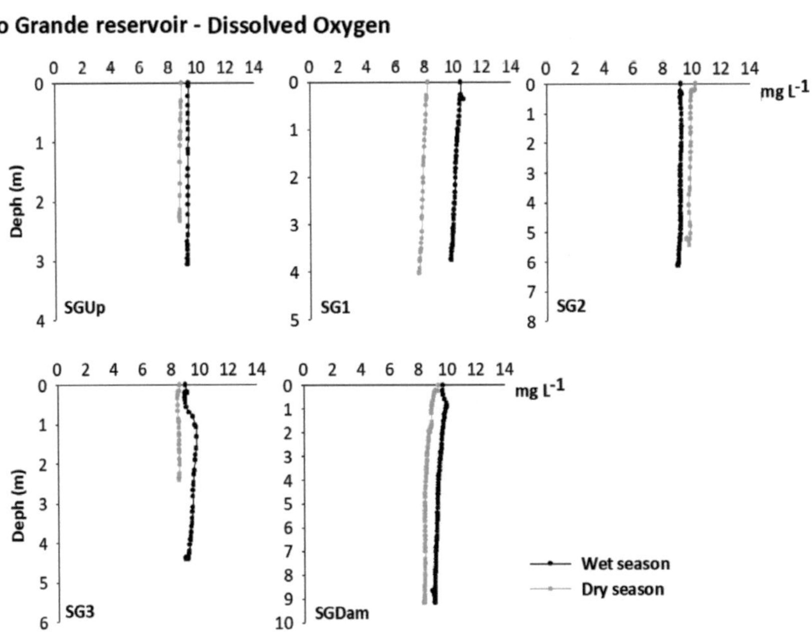

Figure 12. Dissolved oxygen profiles in Salto Grande reservoir sampling sites during wet and dry seasons.

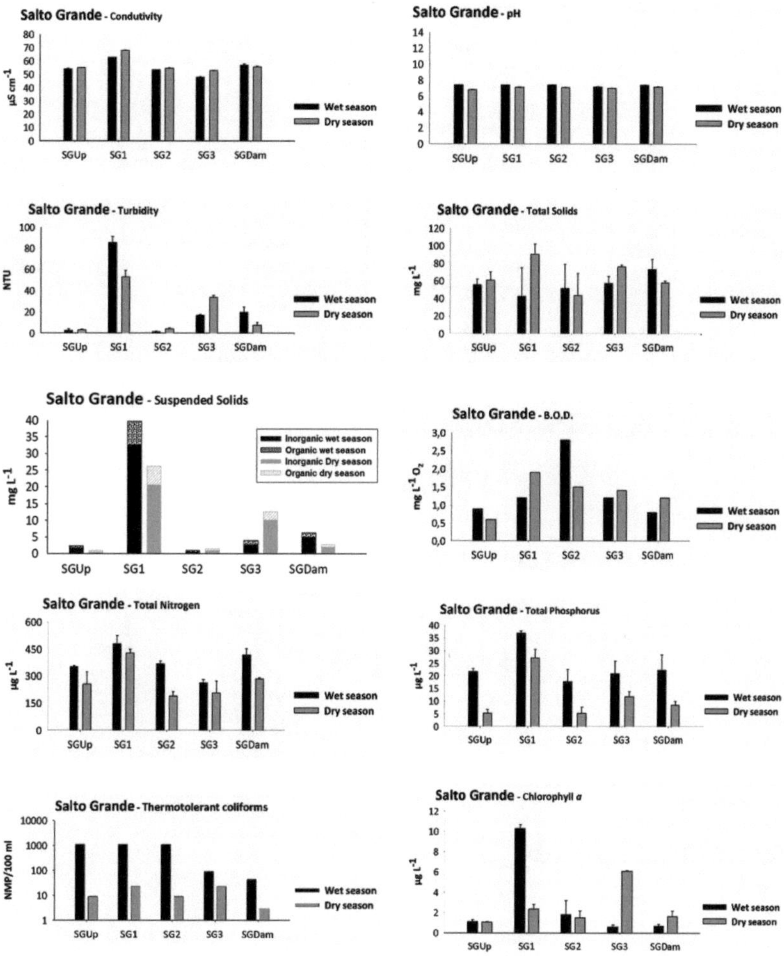

Figure 13. Mean values (+S.D.) of conductivity, pH, turbidity, total and suspended solids, biochemical oxygen demand, total nitrogen, total phosphorus, thermotolerant coliforms and chlorophyll *a* in Salto Grande reservoir sampling sites in the 2011 wet and dry seasons.

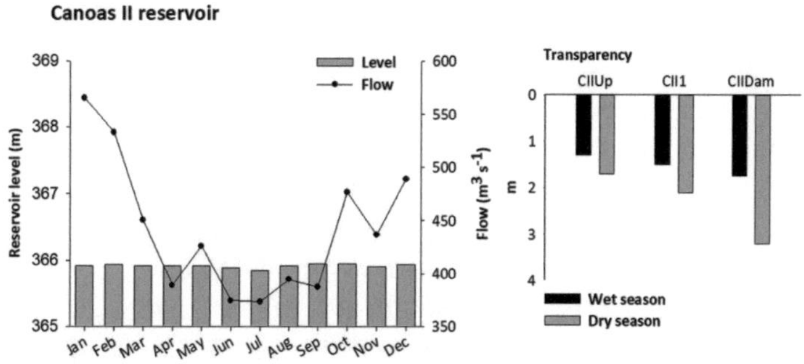

Figure 14. Variation of water level and flow (monthly averages) and transparency (wet and dry seasons) in Canoas II reservoir.

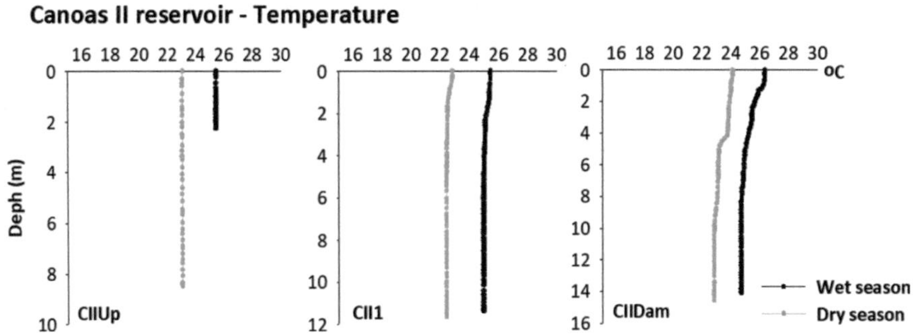

Figure 15. Temperature profiles in Canoas II reservoir sampling sites during wet and dry seasons.

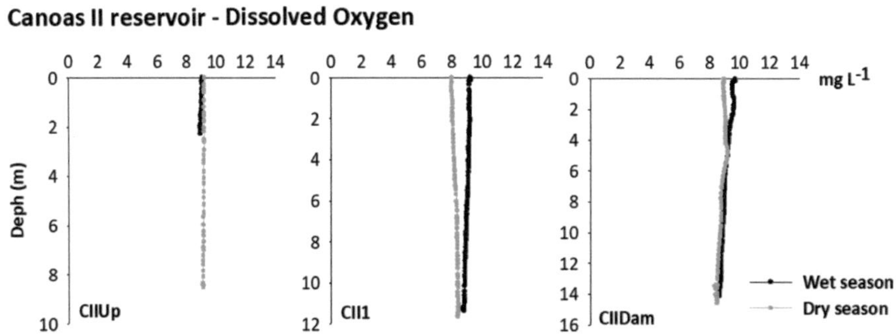

Figure 16. Dissolved oxygen profiles in Canoas II reservoir sampling sites during wet and dry seasons.

5. Inter-reservoir variability

The Paranapanema River reservoir cascade exhibits a wide variability of conditions, when compared the distinct physical dimensions. Salto Grande, the oldest reservoir (60 years), is smaller, with area of 12 km^2; perimeter of 81 km; volume of 63 hm^3; and a very low water retention time, only 2 days. Conversely, the Capivara reservoir (40 years) has the highest area, perimeter, and volume—576 km^2, 1550 km, and 11,743 hm^3, respectively. Capivara is also the most dendritic reservoir, with a shore line development index of 18.2. The two newer (19 years) reservoirs, Canoas I and Canoas II, have the lowest shore line development index, 6.1. The first reservoir in the series, Jurumirim, has the highest water retention time, annual mean of 400 days, followed by the second, Chavantes, with 335 days. The water accumulated in these two upstream reservoirs is fundamental for the flow control along the entire cascade. For both, there was an increase in water release at the end of the dry season (September and October) to supply bellow hydropower plants and, consequently, a decrease in water retention time. In these two storage reservoirs, as well as in Capivara (retention time of 115 days), there is an accentuated fluctuation in the water level, between 2 and 4 m along the studied year. This annual amplitude can be even higher, depending on the year, up to 9 m in Capivara, but not necessarily coupled to the rain/dry regime [17]. In the other reservoirs, operated as run-of-river systems (retention time between 2 and 17 days), the level fluctuation is minimum (<0.5 m). Data show that flow is intensively manipulated in order to keep the best condition for hydroelectric generation. The absolute flow values tend to increase along the river, as expected, from 181 m^3 s^{-1}

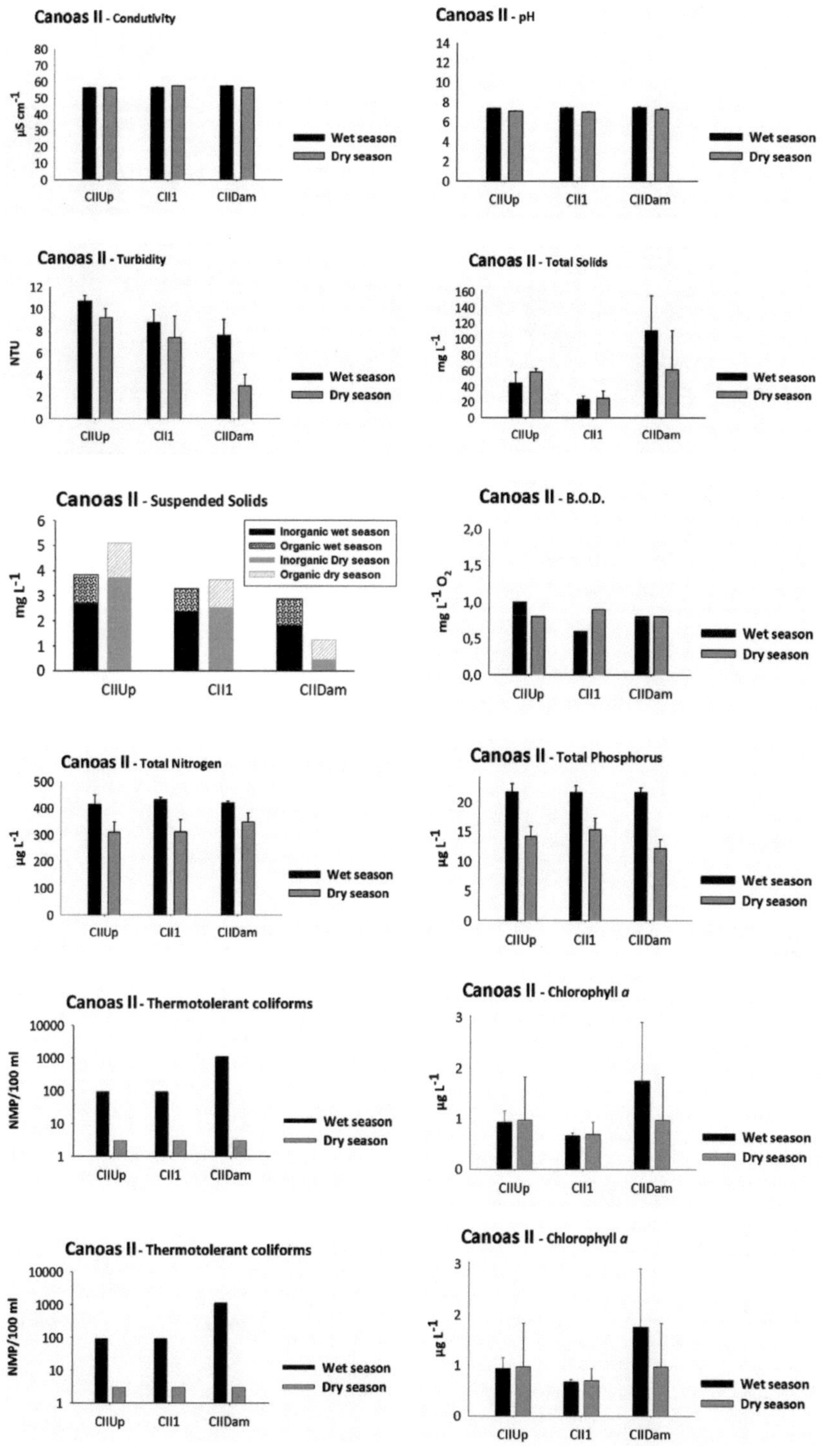

Figure 17. Mean values (+S.D.) of conductivity, pH, turbidity, total and suspended solids, biochemical oxygen demand, total nitrogen, total phosphorus, thermotolerant coliforms and chlorophyll *a* in Canoas II reservoir sampling sites during wet and dry seasons.

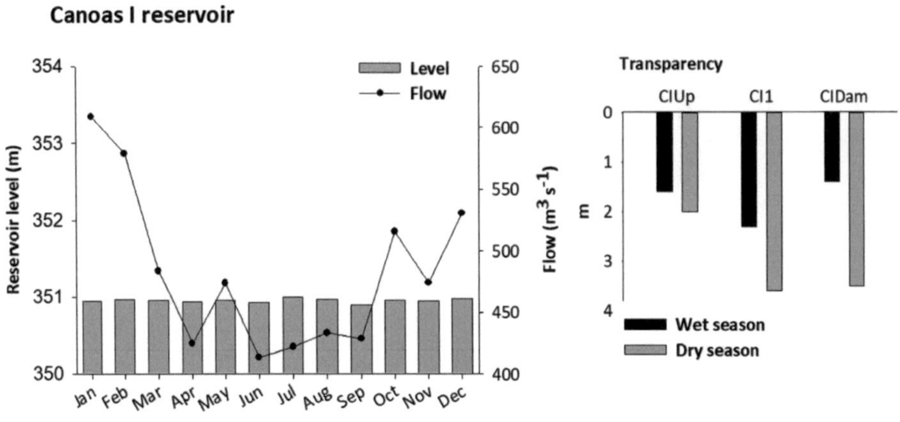

Figure 18. Reservoir level and flow variation throughout the year (monthly average) and water transparency at the Canoas I reservoir sampling sites in the 2011 wet and dry seasons.

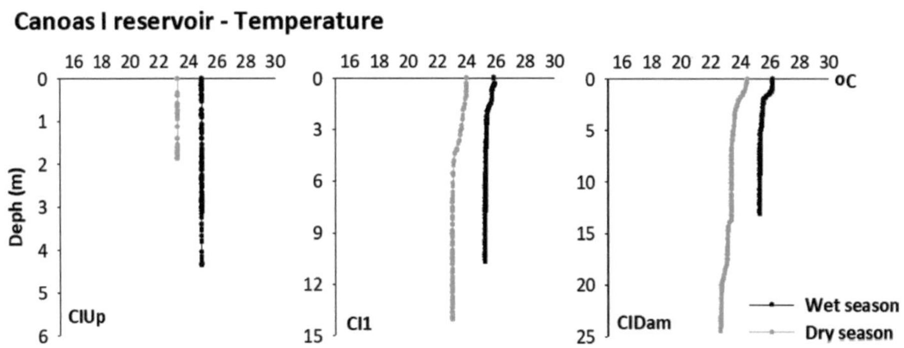

Figure 19. Temperature profiles in Canoas I reservoir sampling sites during wet and dry seasons.

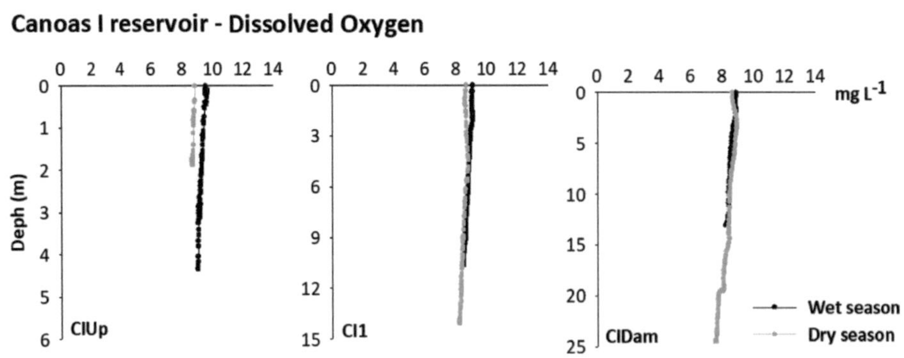

Figure 20. Dissolved oxygen profiles in Canoas I reservoir sampling sites during wet and dry seasons.

registered in Jurumirim (August—dry season) to 1679 $m^3 s^{-1}$ in Rosana (February—rainy season). Ref. [19] analyzed bellow dam releases of Capivara and Taquaruçu reservoirs (ninth and tenth in the cascade) in short periods of time (24 h cycles), showing that differences between

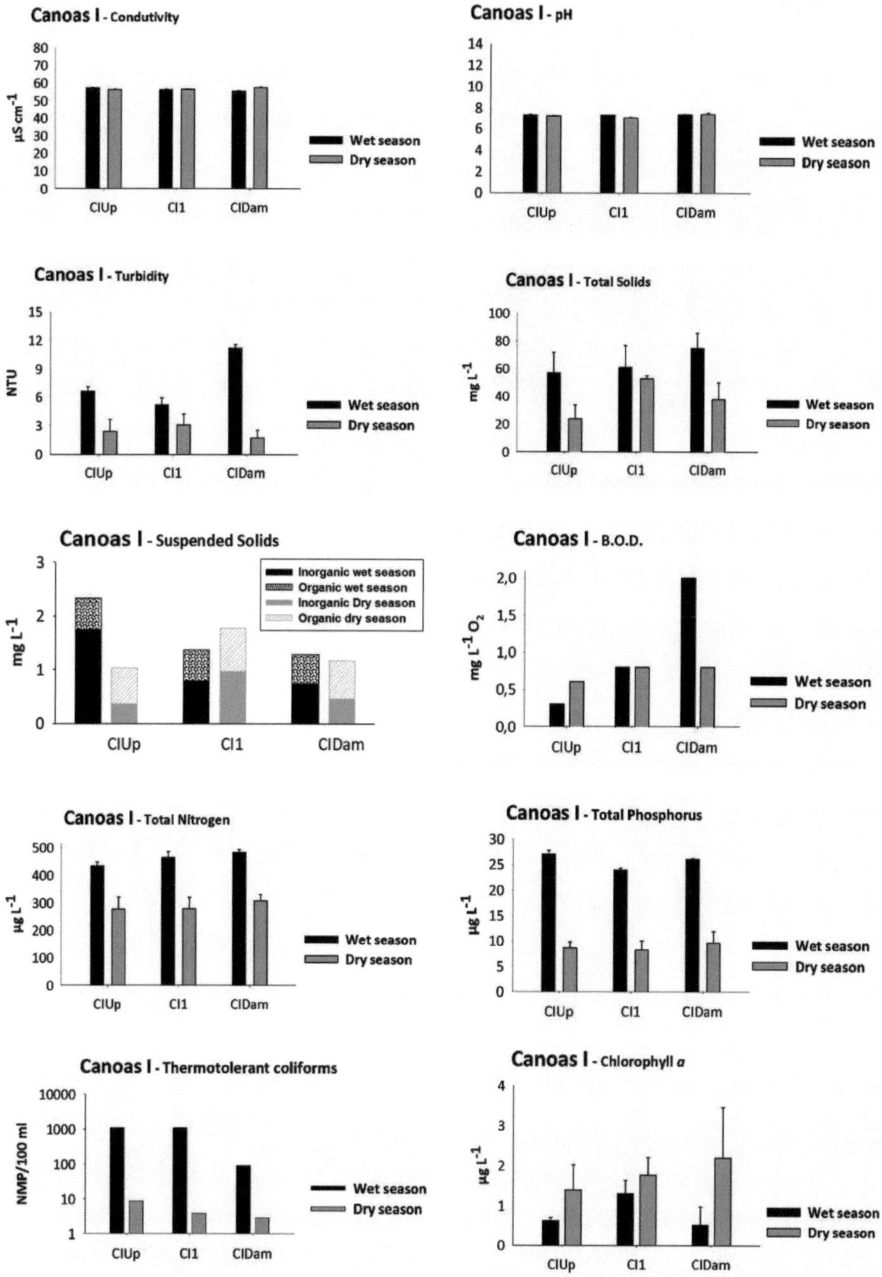

Figure 21. Mean values (mean, +S.D.) of conductivity, pH, turbidity, total and suspended solids, biochemical oxygen demand, total nitrogen, total phosphorus, thermo tolerant coliforms and chlorophyll *a* in Canoas I reservoir sampling sites during wet and dry seasons.

flow peaks reached impressive 1500 $m^3 s^{-1}$. The authors demonstrated that the effects of these hourly discharge variations are significant on the limnological variables' dynamics, especially when intake to the turbines comes from hypolimnion (bottom layers) such as in Capivara

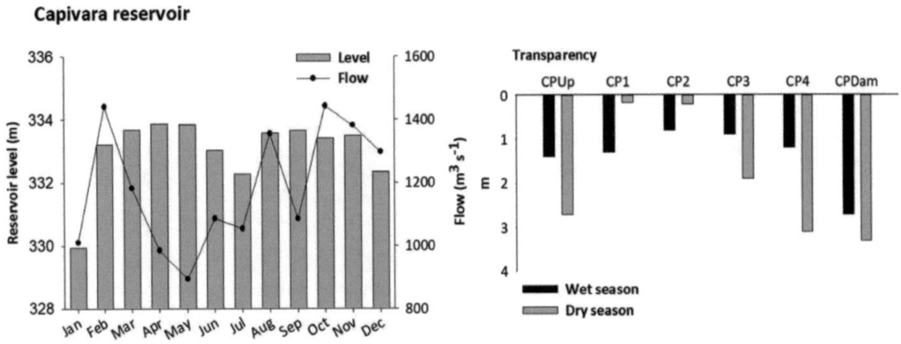

Figure 22. Variation of water level and flow (monthly averages) and transparency (wet and dry seasons) in Capivara reservoir.

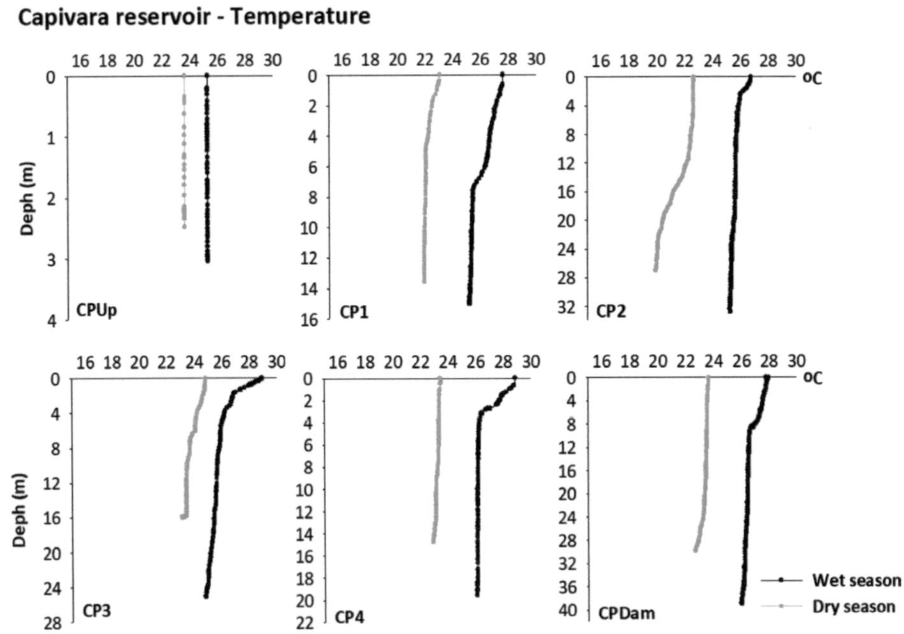

Figure 23. Temperature profiles in Capivara reservoir sampling sites during wet and dry seasons.

dam. In this case, there is a transference of low oxygenated waters, disturbing the limnological and biota rehabilitation of bellow dam river stretches, whose importance has already been evidenced [29–32]. However, this recovery process may not be effective in reservoir cascades, where the distance between consecutive reservoirs is too short. In case of Paranapanema, the longest inter-dam stretch is about 110 km, between Canoas I and Capivara.

The heat accumulation in the reservoir's water mass contributes to an increase of temperature along the cascade. Mean values for the water column varied from 21.4°C (sampling site 12) to 26.9°C (sampling site 33) in wet-summer and from 19.7°C (sampling site 12) to 24.6°C (sampling site 36) in dry-spring. The first corresponds to Chavantes dam, the deepest site in the cascade, and the other value to the two last reservoirs, Taquaruçu and Rosana, respectively.

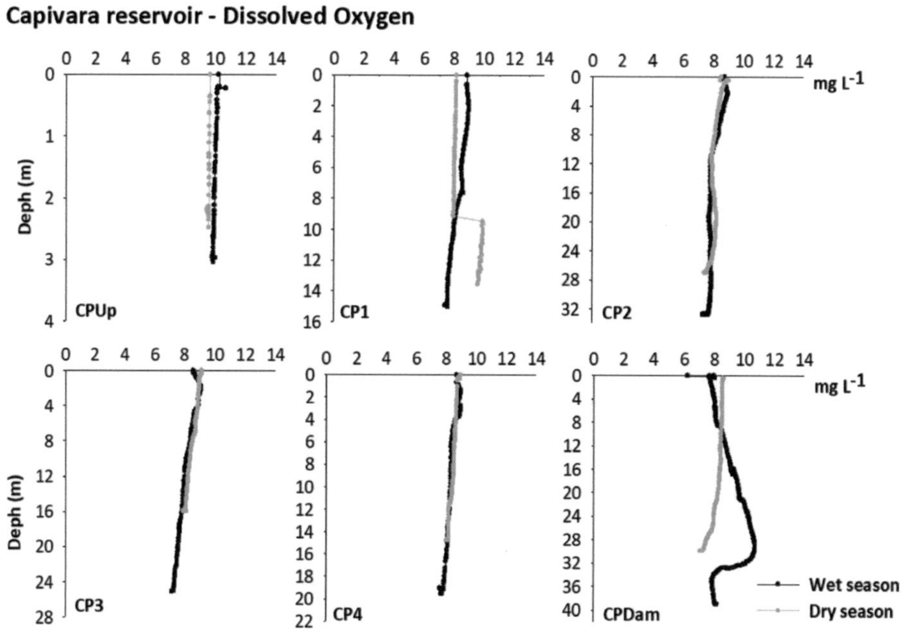

Figure 24. Dissolved oxygen profiles in Capivara reservoir sampling sites during wet and dry seasons.

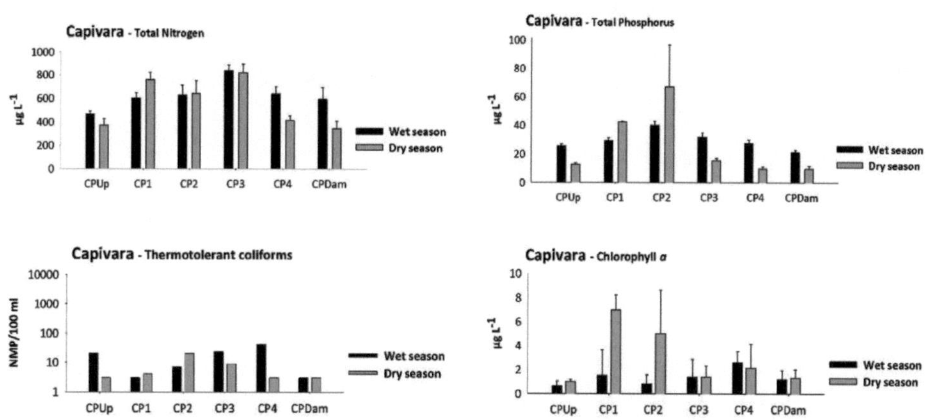

Figure 25. Mean values (+S.D.) of conductivity, pH, turbidity, total and suspended solids, biochemical oxygen demand, total nitrogen, total phosphorus, thermotolerant coliforms and chlorophyll *a* in Capivara reservoir sampling sites during wet and dry seasons.

Principal component analysis, performed to provide an ordination and synthesis of the data set, evidenced the main spatial and temporal trends (**Table 6**; **Figure 34**). The organization during the dry period was mainly represented by Component 1 (explaining 30.7% of data variability) and the wet season by Component 2 (17.6% of data variability). The first component was negatively correlated with transparency and depth, grouping sampling sites (dam zones and central channel sites) of the storage reservoirs: Jurumirim and Chavantes. Sampling stations of Salto Grande, Capivara and Rosana, were associated to the positive side

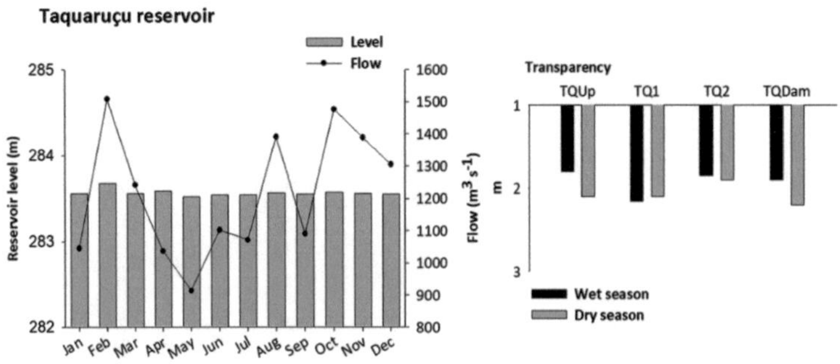

Figure 26. Variation of water level and flow (monthly averages) and transparency (wet and dry seasons) in Taquaruçu reservoir.

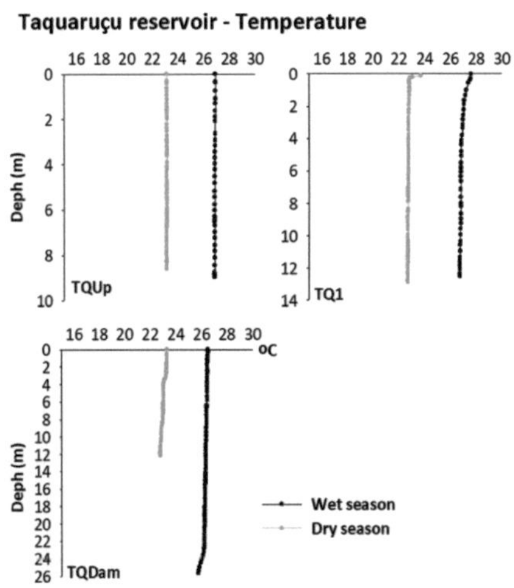

Figure 27. Temperature profiles in Taquaruçu reservoir sampling sites during wet and dry seasons.

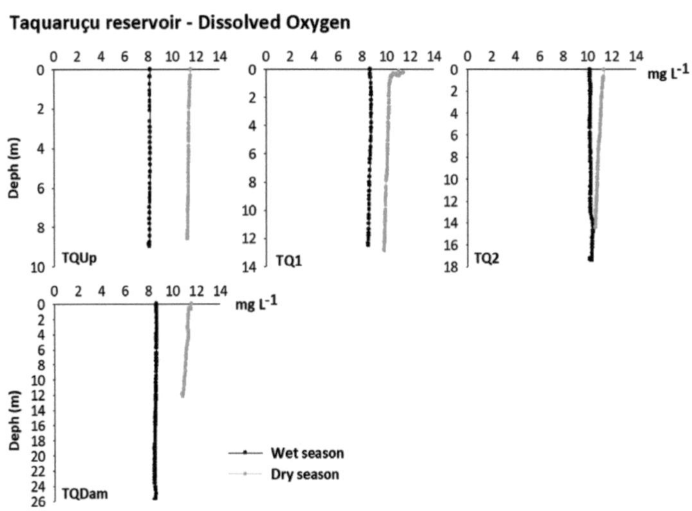

Figure 28. Dissolved oxygen profiles in Taquaruçu reservoir sampling sites during wet and dry seasons.

http://dx.doi.org/10.5772/intechopen.80632

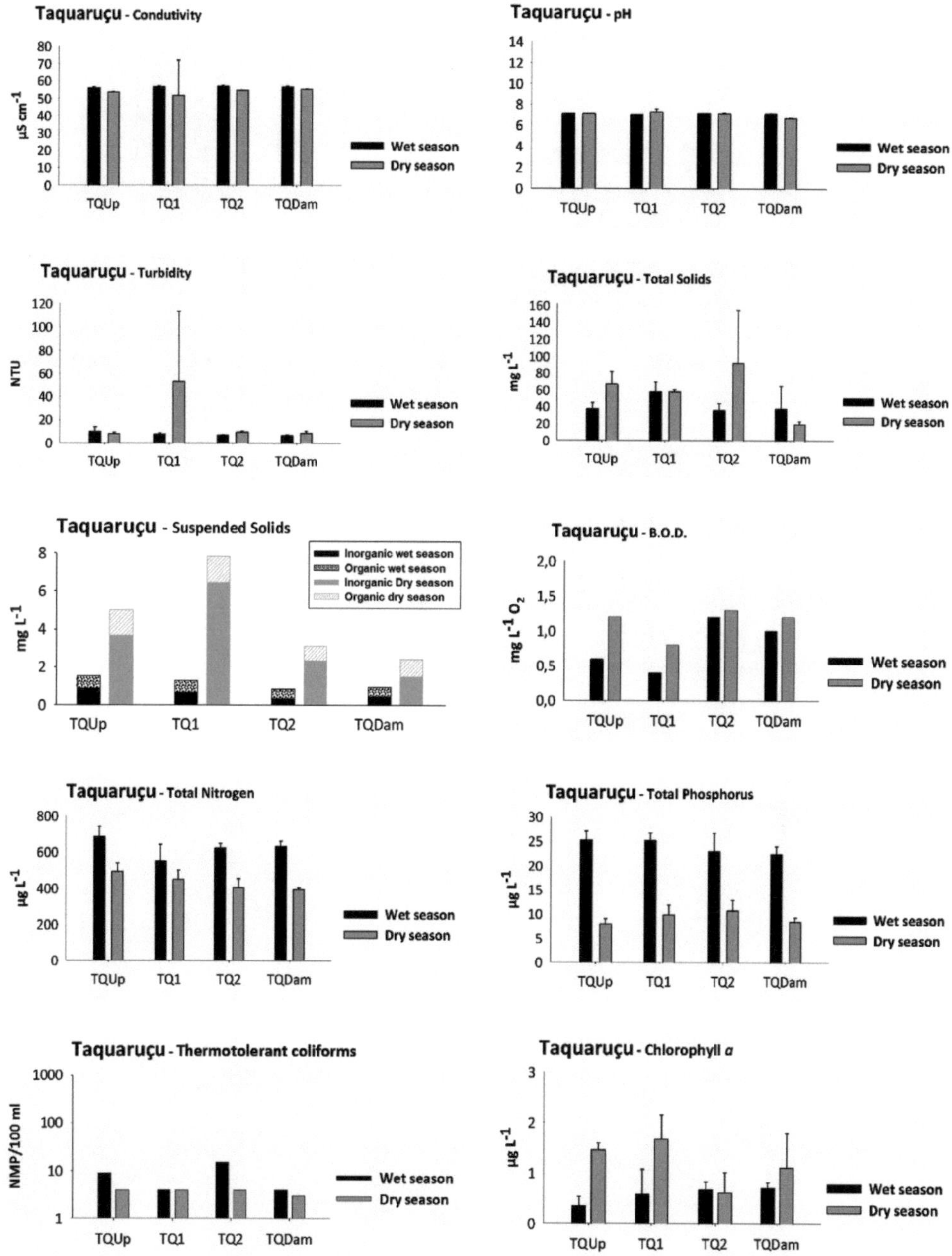

Figure 29. Mean values (mean, +S.D.) of conductivity, pH, turbidity, total and suspended solids, biochemical oxygen demand, total nitrogen, total phosphorus, thermo tolerant coliforms and chlorophyll *a* in Taquaruçu reservoir sampling sites during wet and dry seasons.

of Component 1, determined by suspended solids, nutrients, and chlorophyll (more eutrophic conditions). During the wet season, most sampling stations were associated to the negative side of Component 2, correlated with higher values of temperature, pH, nutrients, and lower concentrations of dissolved oxygen.

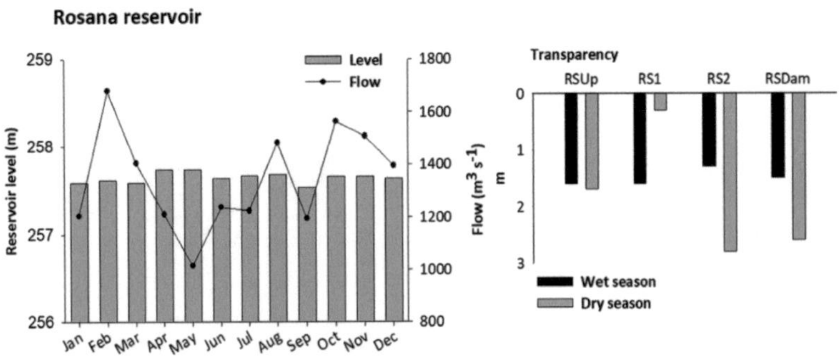

Figure 30. Variation of water level and flow (monthly averages) and transparency (wet and dry seasons) in Rosana reservoir.

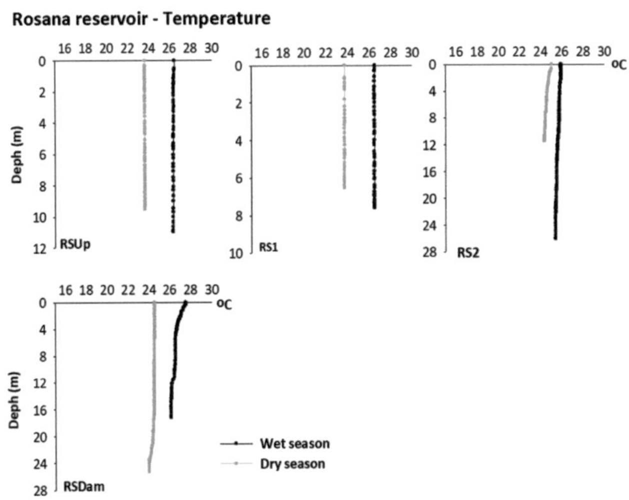

Figure 31. Temperature profiles in Rosana reservoir sampling sites during wet and dry seasons.

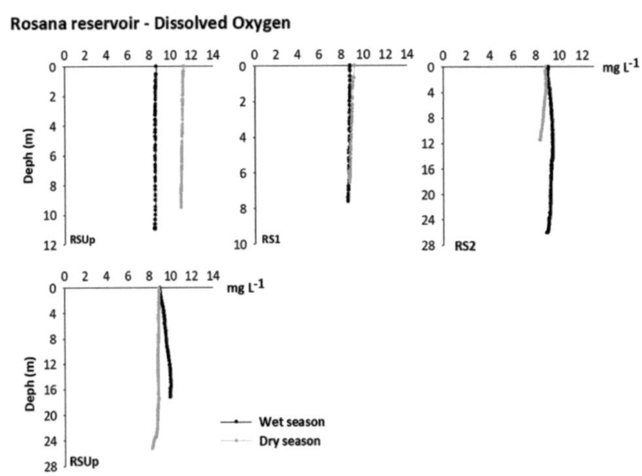

Figure 32. Dissolved oxygen profiles in Rosana reservoir sampling sites during wet and dry seasons.

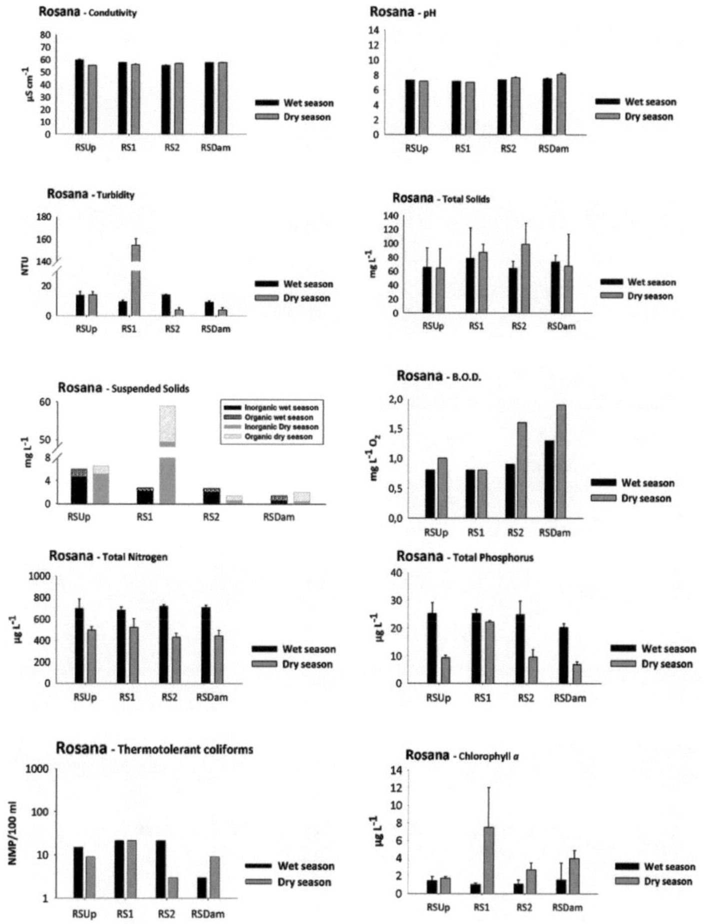

Figure 33. Mean values (mean, + S.D.) of conductivity, pH, turbidity, total and suspended solids, biochemical oxygen demand, total nitrogen, total phosphorus, thermotolerant coliforms and chlorophyll *a* in Rosana reservoir sampling sites during wet and dry seasons.

The predominant trophic state classification for the studied reservoirs was ultraoligotrophic, except for Canoas I, which was oligotrophic. A few results, from a total of 74 individual samples or mean values in case of vertical profiles, were not in conformity with the standard references of the federal legislation for Class 1 waters [28]: 2 values of pH, 6 of turbidity, 1 of chlorophyll, 6 of thermotolerant coliforms, and 19 of total phosphorus. This indicates punctual and isolated sources of degradation. The only exception is phosphorus, above the limit in 25.6% of the measurements. A detailed comparison of the Paranapanema River reservoirs based on distinct water quality and trophic indices is presented by [18]. The study corroborates the good environmental condition of the river but also evidences some negative effects (e.g., phosphorus) associated to the land use intensification—boom of the agrobusiness, especially for the reservoirs in the middle river basin during the wet season [17].

Another potential eutrophication problem is the expansion of cage-aquaculture [33]. This activity has been intensified in the recent years, mainly in Canoas I and Canoas I reservoirs.

Limnological variables	PC1 (30.7%)	PC2 (17.6%)
Transparency	−0.405	−0.020
Depth	−0.125	0.052
Temperature	0.061	−0.545
pH	0.022	−0.420
Conductivity	0.100	−0.184
Turbidity	0.403	0.095
Dissolved oxygen	−0.048	0.251
Biochemical oxygen demand	0.143	0.124
Total solids	0.199	0.061
Inorganic suspended solids	0.415	0.191
Organic suspended solids	0.390	0.165
Total nitrogen	0.225	−0.372
Total phosphorus	0.284	−0.381
Chlorophyll *a*	0.343	0.184
Thermotolerant coliforms	0.089	−0.130

Table 6. Principal component analysis scores (PC1 and PC2) based on the limnological characteristics of the Paranapanema River reservoir cascade.

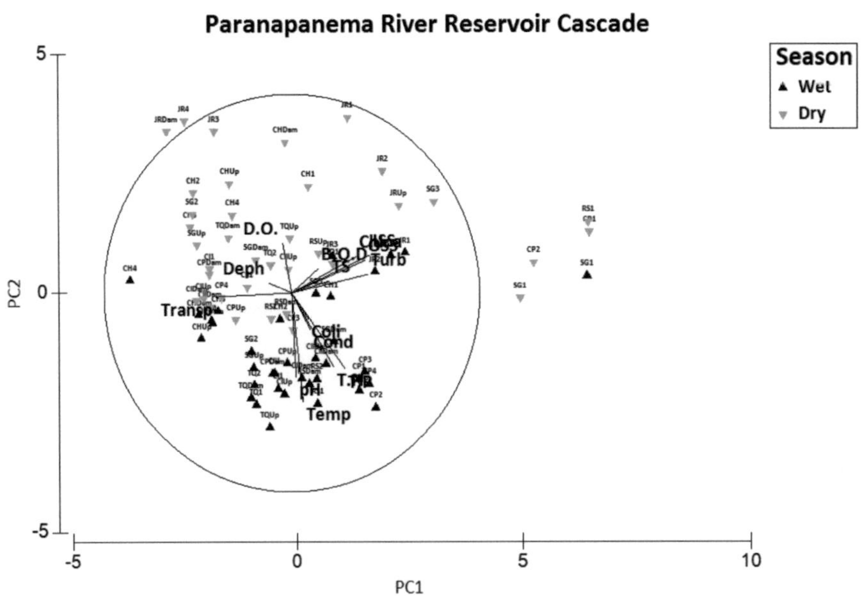

Figure 34. Graphical results of the principal component analysis (PC1 and PC2) of Paranapanema River reservoir cascade based on the limnological variables and using the seasonality as the factor or ordination. For abbreviations see **Tables 1–3**.

6. Intra-reservoir spatial organization—longitudinal dimension

Conspicuous longitudinal gradients (upstream lotic zone → lacustrine dam zone) were observed in the studied reservoirs, especially for the larger storage ones. Classical studies pointed out that compartmentalization enhances the spatial and temporal complexity of reservoirs [5, 13, 34]. This structural characteristic has been verified for several Brazilian reservoirs [7, 10–12, 35, 36].

The Secchi disk transparency, a very simple (easy and cheap) measurement, but limnologically not trivial, is a robust indicator of the presence of distinct water masses along the main axis of a reservoir. In Jurumirim, we registered a clear increasing of transparency toward the dam, about five times in wet season and nine times in dry season. The opposite trend, a longitudinal decrease, occurred for turbidity and suspended solids.

Other variables that indicate well-defined longitudinal gradients were nitrogen, phosphorus, and chlorophyll, whose concentrations reduce toward the dam.

The Chavantes reservoir exhibits an additional spatial complexity due to the existence of "two longitudinal axes," instead of one. An axis corresponds to the former Itararé River Valley (stations CH1, CH2, and CH3), laterally inserted into the Paranapanema River, the other axis (stations CHUp, CH4, and CHDam). In general, variables such as turbidity, suspended solids, total nitrogen, and total phosphorus (in this case only for wet season) exhibited lower values in the Paranapanema axis, compared to the Itararé axis. In both spatial axes, there was a transparency increase toward the dam, with higher values in the Paranapanema, 6.9 and 8.7 times in wet and dry seasons, respectively.

Different from the previously mentioned reservoirs, in Canoas I, Canoas II, Taquaruçu, and Rosana, all run-of-river systems (retention time between 4 and 17 days), the longitudinal gradient was moderate or even a relative homogeneous condition was found. For instance, in Taquaruçu during both campaigns and in Canoas II and Rosana during the wet campaign, the maximum difference along the reservoirs' main axes was only 0.4 m, despite the distance of tens of kilometers between sampling points. For these reservoirs, the magnitude of changes seems to be more relevant on the temporal scale—alternating between dry (winter/early spring) and rainy (late spring/summer) periods. Nutrient concentrations (nitrogen and total phosphorus), for instance, are clearly higher in the wet season.

Very low, and probably transient, transparency in some sampling sites of Rosana and Capivara during the dry season can be attributed to atypical rain events in the upstream areas of both reservoirs immediately (2 days) before sampling. The sediment loads introduced by tributaries (Cinzas River in Capivara and Pirapó River in Rosana) resulted in a remarkable local increase in mineral turbidity (~155 NTU).

The results of Capivara, whose water retention time of 115 days is higher than in run-of-river and lower than in upstream storage reservoirs, demonstrate the importance of both spatial compartmentalization and seasonal variation on the physical and chemical characteristics. The spatial variability is determined by longitudinal gradient and tributary rivers' (more eutrophic waters') contributions. The concentrations of nutrients (nitrogen and phosphorus) varied widely among the different compartments of the reservoir. The main tributary of the Capivara reservoir is the Tibagi River (its mouth corresponds to the sampling site CP3), which drains agricultural areas intensively cultivated and is the second largest river in the whole basin, after the Paranapanema.

In the Salto Grande reservoir, despite its small size, the spatial complexity is high, due to the influence of a large reservoir (Chavantes) located upstream and the entrance of secondary tributary river (Pardo River), which introduces high loads of nutrients and suspended solids [1, 17, 37]. The transparency of water, for example, is higher in the upstream region (SGUp and SG2) and lower at the mouth of the Pardo River (SG1), with a difference of 3.3 m and 2.9 in wet and dry season, respectively. The opposite pattern of variation, as expected, was verified for turbidity and suspended solids, with values lower than 5 NTU at SGUp and SG2 (wet and dry seasons) and higher than 80 NTU in SG1 (wet season).

In addition to the river's prevalent longitudinal pattern of physical, chemical, and biotic organization [38], the results of Capivara and Salto Grande are very important to show that the lateral dimension, mostly the effects of tributaries' entrance, should not be neglected if one aims to understand the spatial structure and functional processes of large rivers and reservoir ecosystems.

7. Intra-reservoir spatial organization: vertical dimension

The development of thermal stratification, commonly followed by chemical differences, is a profound modification in riverine ecosystems after damming. Interactions with the atmosphere are enhanced due to the increase of the exposed surface area, higher water retention time, and reduction in advective transport of mass and energy. In the Paranapanema reservoir cascade, we observed a strong thermal stratification in Chavantes, the deepest reservoir, during summer. Maximum difference between epilimnion (25°C) and hypolimnion (18°C) layers, of 7°C, was verified at the sampling point near the dam (CH12). Other two sampling sites, CH3 and CH4, in the central channel also exhibited a well-defined thermal stratification. The oxygen along the water column decreased approximately from 3 to 4 mg L^{-1}.

During summer, we also observed stratification at the dam zone of the other two storage reservoirs, Jurumirim and Capivara, with differences between layers of 3 and 2°C, respectively. In Jurumirim, particularly, there was a remarkable drop (about 5 mg L^{-1}) of dissolved oxygen in deeper layers, certainly the effect of an extended period (late spring/summer) of stratification. The upstream compartments of these larger reservoirs (especially CHUp, CPUp) were characterized by homogeneous physical and chemical profiles, indicating a continuous mixture regime.

Surficial stratification, probably ephemeral, due to intense solar heating in summer was observed in some sampling sites (e.g., SGDam, SG3, CP3, CP4).

For the other reservoirs, run-of-river systems, isothermal conditions or a slightly gradual decrease of temperature with depth was observed, resulting in a homogeneous or relatively homogeneous distribution of the other variables measured along the water column.

8. Concluding remarks

As previously mentioned, this study has a "historical" importance, because since then it was established a permanent limnological and water quality monitoring program in the Paranapanema River reservoir cascade. Eight consecutive years of continuous and standardized evaluation produced valuable information, which has subsidized important management actions in this large water basin of southeast Brazil.

The larger (and dendritic) storage reservoirs exhibit a well-marked longitudinal compartmentalization, considerable water level fluctuation as well as summer thermal stratification in the central channel and lacustrine zones toward the dam, followed by oxygen decrease in bottom layers. Conversely, run-of-river reservoirs are morphometrically simpler, the spatial complexity is moderate with minor variation in the water level and a continuous mixing regime. Nevertheless, the second kind of reservoirs is less resilient to seasonal changes (dry-wet periods).

In addition to the expected upstream-dam gradients, the lateral component of the spatial heterogeneity was very important for some reservoirs, determined basically by the entrance of tributary rivers which transport considerable loads of nutrients and sediments from intensively cultivated lands.

Finally, it is important to highlight that most measurements were in conformity with the federal standard references for Class 1 waters (good water quality). Exceptions are locally restricted. This fact is corroborated by the predominant low trophic state (77% of ultraoligotrophic determinations). The Paranapanema River reservoir cascade is a strategic regional hydric resource. This is particularly important for the state of São Paulo, the most populous and industrialized state of the country, where important fluvial systems are hypereutrophic and heavily polluted [39] and the metropolitan area of São Paulo city is already facing recurrent water crises, in quantity and quality.

Acknowledgements

The authors are grateful to Dr. Norberto Castro Vianna, from Duke Energy, Geração Paranapanema, the company in charge of hydropower generation in the Paranapanema reservoir cascade (presently China Three Gorges, Brazil) that provides financial support and technical partnership. We also thank the graduate and undergraduate students that kindly helped during field work expeditions and laboratory analyses and the Foundation of the Biosciences Institute—State University of São Paulo (FUNDIBIO) for administrative support (contract 0100015003).

Conflict of interest

None.

Appendix 1: Limnological variables (mean values for the water column) measured in the Paranapanema River reservoir cascade during the wet season

Wet season	Secchi (m)	Deph (m)	Temp. (°C)	pH	Cond. ($\mu S\ cm^{-1}$)	Turb. (NTU)	DO ($mg\ L^{-1}$)	BOD ($mg\ L^{-1}\ O_2$)	TS ($mg\ L^{-1}$)	Inorg. SS. ($mg\ L^{-1}$)	Org. SS ($mg\ L^{-1}$)	TN ($\mu g\ L^{-1}$)	TP ($\mu g\ L^{-1}$)	Chl. *a* ($\mu g\ L^{-1}$)	Thermo Coli (NMP/100 mL)
JR1	0.6	5.1	24.1	7.1	45.5	26.0	9.1	1.3	53.3	8.5	2.6	382.7	20.7	2.6	93
JRUp	0.6	15.0	23.7	7.2	44.4	43.0	8.7	1.0	75.0	6.9	1.7	424.5	24.1	2.6	4
JR2	0.9	18.0	24.1	7.4	60.6	25.5	9.4	1.2	97.0	4.6	1.6	288.9	17.0	2.9	4
JR3	1.0	23.3	23.5	7.4	49.8	17.3	8.8	1.6	89.7	3.3	1.0	349.4	13.3	2.3	4
JR4	2.3	28.8	24.3	7.3	53.6	3.7	8.7	0.2	71.0	0.7	0.9	304.6	11.2	0.8	15
JRDam	3.0	32.0	24.8	7.6	52.3	1.7	8.6	0.8	68.3	0.7	1.2	283.2	13.6	1.0	11
CH1	0.8	17.8	23.7	7.2	54.9	28.5	8.7	0.9	36.0	4.4	1.3	382.6	19.0	1.1	<3
CH2	1.7	24.0	24.3	7.6	53.2	14.8	9.2	1.2	46.0	1.5	1.0	406.0	16.6	1.0	<3
CH3	3.8	48.0	22.6	7.4	54.1	7.0	8.3	1.3	56.3	1.0	0.7	393.0	17.5	0.4	4
CHUp	2.7	10.0	25.4	7.6	50.7	2.0	10.1	0.5	32.7	1.0	0.8	303.3	14.2	0.9	43
CH4	3.7	45.0	22.3	7.2	51.0	2.5	9.6	0.8	11.3	0.4	0.5	301.9	12.8	0.5	4
CHDam	5.5	74.4	21.4	7.2	51.6	5.4	7.7	1.2	35.3	0.6	0.9	372.2	39.8	0.5	4
SGUp	3.5	3.5	25.5	7.4	54.2	2.3	9.4	0.9	55.3	1.6	0.7	349.9	21.6	1.1	>1100
SG1	0.2	4.4	24.2	7.4	62.7	85.4	10.2	1.2	42.3	32.5	7.1	480.5	36.9	10.3	>1100
SG2	3.5	6.4	25.7	7.4	53.3	1.4	9.2	2.8	51.3	0.5	0.6	367.0	17.7	1.8	>1100
SG3	0.8	4.9	25.0	7.2	47.9	16.7	9.2	1.2	57.0	2.7	1.3	263.4	20.9	0.6	93
SGDam	1.6	10.2	25.1	7.4	57.1	19.6	9.4	0.8	73.0	5.0	1.3	418.8	22.4	0.7	43
CIIUp	1.3	8.0	25.5	7.4	56.2	10.7	9.0	1.0	44.7	2.7	1.1	414.4	21.8	0.9	93
CII1	1.5	12.4	25.1	7.4	56.5	8.8	9.0	0.6	23.0	2.4	0.9	432.2	21.7	0.7	93
CIIDam	1.7	13.0	25.2	7.5	57.4	7.6	9.1	0.8	110.7	1.8	1.0	418.5	21.7	1.8	>1100

http://dx.doi.org/10.5772/intechopen.80632

Wet season	Secchi (m)	Deph (m)	Temp. (°C)	pH	Cond. (µS cm⁻¹)	Turb. (NTU)	DO (mg L⁻¹)	BOD (mg L⁻¹ O₂)	TS (mg L⁻¹)	Inorg. SS. (mg L⁻¹)	Org. SS (mg L⁻¹)	TN (µg L⁻¹)	TP (µg L⁻¹)	Chl. *a* (µg L⁻¹)	Thermo Coli (NMP/100 mL)
CIUp	1.6	4.0	24.9	7.3	57.3	6.7	9.3	0.3	57.0	1.7	0.6	434.0	27.1	0.6	>1100
CI1	2.3	9.0	25.4	7.3	56.1	5.2	8.8	0.8	61.0	0.8	0.6	463.8	23.9	1.3	>1100
CIDam	1.4	14.2	25.5	7.4	55.5	11.2	8.6	2.0	74.7	0.7	0.5	485.0	26.1	0.5	93
CPUp	1.4	3.0	25.2	7.3	55.3	9.8	10.0	1.2	69.7	0.4	0.5	468.5	25.5	0.6	21
CP1	1.3	16.0	25.6	7.1	63.9	14.2	8.0	1.2	81.7	3.0	1.0	602.2	29.3	1.5	<3
CP2	0.8	33.0	25.9	7.4	65.2	25.2	8.2	1.3	79.7	3.4	0.9	626.1	40.2	0.8	7
CP3	0.9	26.0	25.9	7.0	48.2	18.8	8.1	0.6	94.3	2.9	1.3	834.7	31.9	1.4	23
CP4	1.2	20.0	26.4	7.3	56.1	12.6	8.3	0.4	104.3	4.2	1.2	638.7	27.7	2.6	43
CPDam	2.7	40	26.6	7.2	55.5	8.9	8.9	0.8	92.0	1.1	0.6	592.9	21.1	1.2	<3
TQUp	1.8	9.5	26.8	7.1	56.4	9.8	8.1	0.6	37.7	0.9	0.6	686.3	25.3	0.3	9
TQ1	2.1	13.0	26.9	7.1	56.9	7.7	8.6	0.4	58.3	0.7	0.6	551.9	25.2	0.6	4
TQ2	1.8	18.0	24.4	7.2	57.0	6.6	10.3	1.2	36.0	0.3	0.5	624.3	23.0	0.7	15
TQDam	1.9	26	26.3	7.2	56.9	6.4	8.6	1.0	38.0	0.5	0.5	634.4	22.4	0.7	4
RSUp	1.6	11.0	26.4	7.3	60.0	13.7	8.6	0.8	65.3	4.6	1.4	696.4	25.2	1.4	15
RS1	1.6	8.5	26.5	7.2	57.6	9.4	8.7	0.8	78.7	2.2	0.6	683.1	25.3	1.0	21
RS2	1.3	27.0	25.6	7.3	55.4	14.0	9.2	0.9	64.0	2.0	0.6	716.0	24.9	1.1	21
RSDam	1.5	18.0	26.5	7.6	57.7	9.0	9.6	1.3	74.0	0.5	0.9	707.2	20.3	1.5	<3

Values not in conformity with the Class 1 standard references in gray.

Appendix 2: Limnological variables (mean values for the water column) measured in the Paranapanema River reservoir cascade during the dry season

Dry season	Secchi (m)	Deph (m)	Temp. (°C)	pH	Cond. (µS cm⁻¹)	Turb. (NTU)	DO (mg L⁻¹)	BOD (mg L⁻¹ O_2)	TS. (mg L⁻¹)	Inorg. SS (mg L⁻¹)	Org. SS (mg L⁻¹)	TN (µg L⁻¹)	TP (µg L⁻¹)	Chl. *a* (µg L⁻¹)	Thermo Coli (NMP/100 mL)
JR1	0.4	4.7	20.8	5.1	51.2	3.8	11.3	1.6	67.3	8.2	1.6	404.3	17.3	1.6	9
JRUp	0.5	13.0	21.5	6.5	57.3	33.1	11.4	1.6	85.7	5.7	1.2	604.6	17.6	2.6	15
JR2	0.6	18.0	21.5	6.4	50.4	36.3	10.9	0.8	89.3	5.7	1.6	404.8	17.6	3.5	11
JR3	2.4	25.0	21.2	6.0	53.9	32.1	11.3	0.8	78.3	1.0	0.6	217.3	4.1	0.7	23
JR4	3.4	25.0	21.1	4.6	56.3	5.1	10.3	0.6	51.7	0.9	0.9	259.4	6.5	0.9	23
JRDam	3.5	29.0	21.3	6.1	51.9	3.7	11.2	1.1	62.7	0.7	0.6	214.7	3.1	1.4	43
CH1	0.9	15.0	21.0	6.6	55.3	21.6	9.8	1.2	77.3	3.7	0.9	302.0	9.0	1.2	15
CH2	3.2	23.0	21.0	6.6	56.1	5.4	10.8	0.6	61.3	0.9	0.6	271.7	6.2	1.3	21
CH3	4.1	40.0	20.5	7.1	55.6	5.4	7.8	1.6	66.7	1.2	0.5	264.1	6.0	0.7	9
CHUp	2.6	16.0	22.0	6.7	50.8	7.2	10.2	0.8	63.3	2.4	1.0	249.4	5.9	1.3	23
CH4	1.4	37.8	20.9	7.1	51.1	13.6	7.5	1.1	68.3	2.3	0.5	197.9	8.8	0.5	3
CHDam	4.2	79.2	19.7	6.9	54.1	5.6	7.5	1.5	78.7	12.7	3.1	264.0	5.3	2.0	4
SGUp	2.3	2.3	22.4	6.8	54.8	3.1	8.8	0.6	60.7	0.6	0.5	254.2	5.2	1.0	9
SG1	0.5	4.0	24.2	7.1	67.9	53.2	7.8	1.9	90.3	20.5	5.7	428.0	27.2	2.3	23
SG2	3.4	5.8	22.7	7.1	54.6	3.9	9.9	1.5	43.7	1.0	0.5	190.8	5.2	1.5	9
SG3	0.4	2.7	23.9	7.0	52.9	33.6	8.5	1.4	76.0	10.1	2.5	206.9	11.8	6.1	23
SGDam	2.4	10.7	23.3	7.1	55.8	7.7	8.7	1.2	57.7	2.0	0.8	283.1	8.5	1.6	<3
CIIUp	1.7	9.5	23.0	7.1	56.2	9.2	9.1	0.8	58.7	3.7	1.4	308.8	14.1	1.0	<3
CII1	2.1	12.6	22.5	7.0	57.4	7.4	8.3	0.9	24.7	2.5	1.1	310.0	15.3	0.7	<3
CIIDam	3.2	15.5	23.2	7.3	56.4	3.0	8.8	0.8	61.7	0.4	0.8	346.8	12.1	1.0	<3

Dry season	Secchi (m)	Deph (m)	Temp. (°C)	pH	Cond. (μS cm^{-1})	Turb. (NTU)	DO (mg L^{-1})	BOD (mg L^{-1} O_2)	TS. (mg L^{-1})	Inorg. SS (mg L^{-1})	Org. SS (mg L^{-1})	TN (μg L^{-1})	TP (μg L^{-1})	Chl. *a* (μg L^{-1})	Thermo Coli (NMP/100 mL)
CIUp	2.0	2.0	23.2	7.2	56.3	2.4	8.9	0.6	23.7	0.4	0.7	278.6	8.6	1.4	9
CI1	3.6	14.9	23.2	7.0	56.6	3.1	8.5	0.8	52.7	1.0	0.8	279.0	8.2	1.8	4
CIDam	3.5	25.2	23.2	7.4	57.5	1.7	8.3	0.8	37.7	0.5	0.7	306.9	9.7	2.2	<3
CPUp	2.7	2.7	23.5	7.1	57.3	2.8	8.0	1.2	56.0	0.2	0.9	371.4	12.9	1.0	<3
CP1	0.2	12.5	22.1	7.0	55.0	144.5	8.4	1.9	113.7	39.7	4.5	760.7	42.4	7.0	4
CP2	0.2	27.3	21.3	6.8	58.3	88.7	8.1	1.6	113.7	17.3	2.6	640.7	67.2	5.0	20
CP3	1.9	16.5	23.8	7.1	48.3	11.0	8.5	0.5	74.0	1.7	0.9	818.7	15.6	1.4	9
CP4	3.1	16.8	23.2	7.3	54.0	3.2	8.4	0.8	57.3	0.5	0.6	413.8	9.8	2.2	<3
CPDam	3.3	29.5	23.3	7.2	53.1	2.8	9.6	0.4	174.7	0.3	0.8	343.2	9.8	1.3	<3
TQUp	2.1	9.5	23.1	7.2	53.8	7.9	11.3	1.2	67.0	3.7	1.3	494.8	8.0	1.5	4
TQ1	2.1	12.8	23.0	7.3	51.7	53.0	8.5	0.8	58.7	6.5	1.4	450.0	9.9	1.7	4
TQ2	1.9	15.0	22.6	7.2	54.8	9.5	9.4	1.3	92.7	2.4	0.8	403.9	10.8	0.6	4
TQDam	2.2	16.0	23.0	6.8	55.5	8.3	11.1	1.2	19.7	1.5	0.9	395.3	8.5	1.1	<3
RSUp	1.7	11.9	23.6	7.2	55.3	13.9	11.1	1.0	64.3	5.1	1.4	498.6	9.3	1.7	9
RS1	0.3	7.0	23.7	7.0	56.3	154.8	8.9	0.8	87.3	49.5	9.4	521.8	22.2	7.5	21
RS2	2.8	13.4	24.6	7.7	57.0	4.0	8.7	1.6	99.3	0.6	0.8	429.2	9.5	2.7	<3
RSDam	2.6	25.9	24.4	8.1	57.7	3.9	8.7	1.9	67.7	0.4	1.6	443.6	6.8	4.0	9

Values not in conformity with the Class 1 standard references in gray.

Author details

Marcos Gomes Nogueira* and Juliana Pomari

*Address all correspondence to: marcos.nogueira@unesp.br

State University of São Paulo (UNESP), Botucatu, São Paulo, Brazil

References

[1] Zarlf C, Lumsdon AE, Berlekamp J, Tydecks L, Tockner K. A global boom in hydropower dam construction. Aquatic Sciences. 2015;**77**:161-170. DOI: 10.1007/s00027-014-0377-0

[2] Nogueira MG, Perbiche-Neves G, Naliato DAO. Limnology of two contrasting hydroelectric reservoirs (storage and run-of-river) in southeast Brazil. In: Samadi-Boroujeni H, editor. Hydropower. Rijeka, Croatia: Intech; 2012. pp. 167-184

[3] Rodrigues SC, Silva TI. Dam construction and loss of geodiversity in the Araguari River basin, Brazil. Land Degradation and Development. 2012;**23**:419-426. DOI: 10.1002/ldr.21

[4] Brambilla EM, Ruocco AMC, Nogueira MG. A contribution for the limnological knowledge of basaltic knickones. Brazilian Journal of Biology. 2018;**78**(2):375-385. DOI: 10.1590/1519-6984.172985

[5] Thornton WK. Perspectives on reservoir limnology. In: Thornton KW, Kimmel BL, Payne EF, editors. Reservoir Limnology: Ecological Perspectives. New York: John Wiley and Sons; 1990. pp. 1-13

[6] Wetzel RG. Limnology: Lake and River Ecosystems. 3rd ed. San Diego: Academic Press; 2001. 1006 p

[7] Nogueira MG, Henry R, Maricatto FE. Spatial and temporal heterogeneity in the Jurumirim Reservoir, São Paulo, Brazil. Lakes and Reservoirs: Research and Management. 1999;**4**:107-120. DOI: 10.1046/j.1440-1770.1999.00086.x

[8] Nogueira MG. Phytoplankton composition, dominance and abundance as indicators of environmental compartmentalization in Jurumirim Reservoir (Paranapanema River), São Paulo, Brazil. Hydrobiologia. 2000;**431**:115-128. DOI: 10.1023/A:1003769408757

[9] Nogueira MG, Ferrareze M, Moreira ML, Gouvêa RM. Phytoplankton assemblages in a reservoir cascade of a large tropical–subtropical river (SE, Brazil). Brazilian Journal of Biology. 2010;**70**(3):781-793. DOI: 10.1007/s10750-004-2552-0

[10] Pinto-Coelho RM, Azevedo LM, Rizzi PEV, Bezerra-Neto JF, Rolla ME. Origens e efeitos do aporte externo de nutrientes em um reservatório tropical de grande porte: Reservatório de São Simão (MG/GO). In: Nogueira MG, Henry R, Jorcin A, editors. Ecologia de reservatórios: Impactos potenciais, ações de manejo e sistemas em cascata. 2nd ed. São Carlos: Rima; 2006. pp. 127-164

[11] Henry R, Maricatto FE. Sedimentation rates of tripton in Jurumirim reservoir (São Paulo, Brasil). Limnologica. 1996;**25**:15-25

[12] Pagioro TA, Thomaz SM. Longitudinal patterns of sedimentation in a deep, monomitic subtropical reservoir (Itaipu, Brazil–Paraguay). Archives of Hydrobiology. 2002;**154**(3): 515-528. DOI: 10.1127/archiv-hydrobiol/154/2002/515

[13] Armengol J, Garcia JC, Comerma M, Romero M, Dolzs J, Roura M, et al. Longitudinal processes in canyon type reservoirs: The case of Sau (N.E. Spain). In: Tundisi JG, Straškraba M, editors. Theoretical Reservoir Ecology and its Applications. Leiden: Brazilian Academy of Sciences, International Institute of Ecology and Backhuys Publishers; 1999. pp. 313-345

[14] Straškraba M. Coupling of hydrobiology and hydrodynamics: Lakes and reservoirs. In: Physical Processes in Lakes and Reservoirs, Coastal and Estuarine Studies. Vol. 54. Washington, DC: American Geophysical Union; 2013. pp. 601-622. DOI: 10.1029/CE054p0623

[15] Tundisi JG, Straškraba M, editors. Theoretical Reservoir Ecology and its Applications. Leiden: Brazilian Academy of Sciences, International Institute of Ecology and Backhuys Publishers; 1999. 585 p

[16] Tundisi JG, Matsumura-Tundisi T. Integration of research and management in optimizing multiple uses of reservoirs: The experience of South American and Brazilian case studies. Hydrobiologia. 2003;**500**:231-242. DOI: 10.1023/A:102461710205

[17] Nogueira MG, Jorcin A, Vianna NC, Britto YC. Reservatórios em cascata e os efeitos na limnologia e organização das comunidades bióticas (fitoplâncton, zooplâncton e zoobentos): Um estudo de caso no rio Paranapanema (SP/PR). In: Nogueira MG, Henry R, Jorcin A, editors. Ecologia de reservatórios: Impactos potenciais, ações de manejo e sistemas em cascata. 2nd ed. São Carlos: Rima; 2006. pp. 435-459

[18] Pomari J, Kane DD, Nogueira MG. Application of multiple-use indices to assess reservoirs water quality and the use of plankton community data for biomonitoring purposes. International Journal of Hydrology. 2018;**2**(2):168-179

[19] Valderrama JG. The simultaneous analysis of total nitrogen and phosphorus in natural waters. Marine Chemistry. 1981;**10**:109-122

[20] Mackereth FIH, Heron J, Talling JF. Water analysis: Some revised methods for limnologists. Freshwater Biological Association Scientific Publication No. 36. Second Impression; 1989. 120 p

[21] Strickland JD, Parsons TR. A manual of sea water analysis. Fisheries Research Board of Canada. 1960;**125**:1-185

[22] American Public Health Association. Standard Methods for the Examination of Water and Wastewater. 22nd ed. Washington: American Public Health Association; 2012. 1496 p

[23] Wetzel RG, Likens GE. Limnological Analyses. 2nd ed. New York: Springer-Verlag; 1991. 391 p

[24] Talling JF, Driver D. Some problems in the estimation of chlorophyll *a* in phytoplankton. In: Proceedings, Conference of Primary Productivity Measurements in Marine and Freshwater. Hawaii: USAEE; 1963. pp. 142-146

[25] Golterman HL, Clymo RS, Ohstad MA. Methods for Physical and Chemical Analysis of Fresh Waters. 2nd ed. Oxford: Blackwell Science; 1978

[26] Håkanson L. The importance of lake morphometry and catchment characteristics in limnology—ranking based on statistical analyses. Hydrobiologia. 2005;**541**:117-137. DOI: 10.1007/s10750-004-5032-7

[27] Cunha DGF, Calijuri MDC, Lamparelli MC. A trophic state index for tropical/subtropical reservoirs. Ecological Engineering. 2013;**60**:126-134. DOI: 10.1016/j.ecoleng.2013.07.058

[28] CONAMA. Available from: http://www.mma.gov.br/port/conama/res/res05/res35705.pdf [Accessed: June 15, 2018]

[29] Naliato DAO, Nogueira MG, Perbiche-Neves G. Discharge pulses of hydroelectric dams and their effects in the downstream limnological conditions: A case study in a large tropical river (SE Brazil). Lakes and Reservoirs: Research and Management. 2009;**14**: 301-314. DOI: 10.1111/j.1440-1770

[30] Matsuura P, Perbiche-Neves G, Ferreira RAR, Nogueira MG. Changes in the phytoplankton structure downstream a large reservoir: Effects of tributaries on the assemblages attributes. Biologia Section Botany. 2015;**70**(3):320-327. DOI: 10.1515/biolog-2015-0037

[31] Portinho J, Perbiche-Neves G, Nogueira MG. Zooplankton community and tributary effects in free-flowing section downstream a large tropical reservoir. International Review of Hydrobiology. 2016;**10**:1-9. DOI: 10.1002/iroh.201501798

[32] Moura AN, Severiano JS, Tavares NKA, Dantas EW. The role of a cascade of reservoirs and seasonal variation in the phytoplankton structure in a tropical river. Brazilian Journal of Biology. 2013;**73**(2):291-298. DOI: 10.1590/S1519-69842013000200009

[33] Ayroza DMMR, Nogueira MG, Ayroza LMS, Carvalho ED, Ferraudo ADS, Camargo AFM. Temporal and spatial variability of limnological characteristics in areas under the influence of Tilapia cages in the Chavantes Reservoir, Paranapanema River, Brazil. Journal of the World Aquaculture Society. 2013;**44**(6):814-825. DOI: 10.1111/jwas.12082

[34] Straškraba M, Tundisi JG, Duncan A. State-of-the-art of reservoir limnology and water quality management. In: Straškraba M, Tundisi JG, Duncan A, editors. Comparative Reservoir Limnology and Water Quality Management. Dordrecht: Kluwer Academic Publishers; 1993. pp. 213-288

[35] Jorcin A, Nogueira MG. Temporal and spatial patterns along the cascade of reservoirs in the Paranapanema River (SE Brazil) based on the characteristics of sediment and sediment-water interface. Lakes and Reservoirs: Research and Management. 2005;**10**: 1-12. DOI: 10.1111/j.1440-1770

[36] Soares MCS, Marinho MM, Huszar VLM, Branco CWC, Azevedo SMFO. The effects of water retention time and watershed features on the limnology of two tropical reservoirs in Brazil. Lakes and Reservoirs: Research and Management. 2008;**13**:257-269. DOI: 10.1111/j.1440-1770.2008.00379.x

[37] Perbiche-Neves G, Nogueira MG. Reservoir design and operation: Effects on aquatic biota—A case study of planktonic copepods. Hydrobiologia. 2013;**707**(1):187-198. DOI: 10.1007/s10750-012-1425-1

[38] Vannote R, Minshall G, Cummins K, Sedell J, Cushing C. The river continuum concept. Canadian Journal of Fisheries and Aquatic Sciences. 1980;**37**(1):130-137. DOI: 10.1139/f80-017

[39] Tundisi JG, Matsumura-Tundisi T, Abe DS. The ecological dynamics of Barra Bonita (Tietê River, SP, Brazil) reservoir: Implications for its biodiversity. Brazilian Journal of Biology. 2008;**68**(4, Suppl):1079-1098. DOI: 10.1590/S1519-69842008000500015

Section 3

Water Quality of Aquaculture

Chapter 7

Importance of Optimum Water Quality Indices in Successful Frog Culture Practices

Cleber Fernando M. Mansano, Luiz Sérgio Vanzela,
Juliana Heloisa P. Américo-Pinheiro,
Beatrice I. Macente, Kifayat U. Khan,
João Batista K. Fernandes,
Danila Fernanda R. Frias and Marta V. De Stéfani

Additional information is available at the end of the chapter

http://dx.doi.org/10.5772/intechopen.81747

Abstract

The optimum quality of water indices is extremely important for successful frog culture. Frogs excrete their excreta and skin debris in water. Therefore, it is necessary to regularly renew the water and clean the tanks and bays of rearing systems. Such care is necessary for the prevention and prophylaxis of diseases, which may cause severe mortalities. Bullfrogs need water of good physical and chemical quality, and thus, water quality indices must be measured before starting a breeding and rearing program. Additionally, the producers should have a good knowledge about the water quality before establishing a rearing system. Aquatic ecosystems are dynamic and even in small rearing water tanks, physical and chemical parameters are interrelated. For example, any change in dissolved oxygen level depends on the water temperature and atmospheric pressure. The dissolved oxygen level is almost 9.08 mg L^{-1} near sea side at a temperature of 20°C, whereas its concentration rises up to 10.07 mg L^{-1}, if the temperature drops to 15°C, indicating that dissolved oxygen and water temperature are closely interrelated. Thus, physical and chemical parameters of water should be considered and analyzed together because all of these factors have a direct impact on the culture systems.

Keywords: aquatic ecosystems, *Lithobates catesbeianus*, raniculture effluent

1. Introduction

Frog culture consists of small- and medium-scale producers and is gaining the attention since several years. The improvements in production systems and research have contributed a lot to make it more viable and profitable for producers. Brazil is considered as the country that has the best technologies in the production of bullfrog, that is, the entire production cycle is completed under controlled conditions. However, the sector encounters some limitations such as fewer improvements in the areas of nutrition, reproduction and genetics, mismanagement of the sector, and little investment and actions by the public sector [1].

The frog's production has different markets including the domestic and foreign markets. The main product is fresh meat, represented mostly by the trade of frogs' thighs [2]. More than 90% of internationally traded frog meat comes from extractive hunting, but there is enormous pressure from environment protection agencies that are claiming the finishing of such practice. Data published by FAO [3] on world ranching showed Brazil as the second largest producer, the first being Taiwan.

In Brazil and China, raniculture is exclusively based on the production and rearing of bullfrog (*Lithobates catesbeianus*), an exotic species from North America that was introduced for captive breeding. When compared to native species, it presents higher productive performance in commercial productions [4]. The species presents extremely favorable zoo-technical characteristics for high-scale breeding, such as precocity, prolificacy, and rusticity, which greatly facilitate management [2].

As among all amphibian species, frogs are ectothermic animals, that is, the temperature of water and environment directly influences the animal's metabolism [5]. The temperature of water between 22 and 28°C is considered ideal for the excellent development of tadpoles [6] and may result in achieving better zoo-technical indices in the rearing phase [7].

Amphibians like bullfrog need water of specific physical and chemical quality. For producers, before starting a rearing system, it is extremely important to know and has a good knowledge about different water quality parameters and their role in successful frog culture. They should measure the pH, electrical conductivity, total alkalinity, total hardness, ammonia, nitrite, nitrate, phosphorus, chlorides, iron, and especially oxygen properly before starting breeding. These parameters are the most important indices that indicate the water quality.

2. Usage, water quality, and aquaculture effluents

Aquaculture has shown great development in the last decades and is competing with production systems of other aquatic animals due to water resource. Its development, however, presents risks of deteriorating the quality and quantity of water and in contributing declining the environmental, social, and economic quality. Technical, scientific, and representative links of Brazilian aquaculture have stated that frog culture does not consume but rather uses water, and this characteristic of nonconductivity could change approaches and strategies related to the management of water resources directed to aquatic productions, taking in consideration different from the industries [8].

The granting and charging due to the use of water resources by aquaculture become more relevant when it highlights the aspects of water quality used and release the water source at the expense of aspects that basically involve the use of large volumes of water. The usage of aquaculture water depends greatly on water quality [9].

Aquaculture uses water resources intensively, competing for water available to the population and other productive systems. However, unlike other production systems, aquaculture can collaborate with water quality control systems by constant monitoring while keeping in view its role in producing food for humans especially protein food [8].

An objective and consistent model of granting and charging due to the use of water for aquaculture refers to methods focused on quality differentials, which are possibly based on measurements of biochemical oxygen demand (BOD) and levels of nitrogen (N) and phosphorus (P) in the water collected and returned to its source or original course. Other factors, such as water surface area, chosen design, or management techniques, may interfere with higher actual water consumption and may also be taken into account in the models of granting and charging the use of water resources for aquaculture [8].

Evaluating water use in the last 100 years, Telles [10] stated that globally about 70% of the available water is destined for agricultural usage. According to the author, approximately 20% of the water is destined to the industry and less than 10% to population (hygiene and direct consumption). In response to increasing food demands, aquaculture has also advanced its production. In the last decade, world production has increased by 200% mainly searching for a healthy product and the next step is the search for an ecologically correct product.

For any production system to continue its growth, it must be economically and environmentally correct [11, 12]. In addition, it should be based on the concepts of food security and socioeconomic development [13]. According to Valenti et al. [13], modern aquaculture is based on three pillars: profitable production, social development, and preservation of the environment, having an intrinsic and interdependent relationship for a perpetual production.

Although aquaculture provides a number of social and economic benefits, it must seek new technologies to reduce impacts such as deforestation, diversion of watercourses, introduction of exotic species into the natural environment, and effluent emission into aquifers. Environmental pollution can be defined as any action or omission of man, which has a direct impact on water, soil, and air and causes a harmful imbalance on the environment as well [14]. The impacts caused by aquaculture result in reduced production, disease outbreaks in cultivated and wild populations, and in some cases may restrict aquaculture operations [15].

Environmental factors such as soil quality, water quality, risks of introducing exotic species/biodiversity, chemical and organic discharge in the natural environment, recycling and interaction with other neighboring fish farms have a great impact on fish farming [16]. Another type of impact is that which is caused by aquaculture on water quality is related to the accumulation of nutrients and organic residues at the bottom of tanks and nurseries and such impact is time dependent. This time is related to unused food, fertilizer usage, and the nutrients present in the sediment, which precipitate and then release into the water column [17].

The nurseries have high concentrations of nutrients, plankton, organic, and inorganic matter due to the food provided and the morphometry of the systems. The input and output of constant water with short residence time and uncontrolled management of feed and water are some of the important factors that directly act on nutrient dynamics, accelerating or allowing greater availability in the water column [18].

The water inflow and outflow due to its great intensity remove excess nutrients and other material from the nursery, controlling the phosphorus dynamics in the medium in relation to its absorption in the sediment [19]. In times of high fish production, from November to April (increased water temperature), the addition of feed is more intense and climatic factors such as temperature and precipitation influence the dynamics of these systems [18].

The impacts of aquaculture can be classified as internal, local, and/or regional. Internal impacts refer to those that interfere with the breeding system itself, such as the depletion of dissolved oxygen in a fish farm. In general, local impacts extend 1 km downstream of the effluent discharge. Effects on aquatic environments with a spatial scale of several kilometers are considered regional impacts [20].

The effluents from aquaculture ponds have a high volume and low nutrient concentrations when compared to domestic effluents, which present little volume and high concentration of nitrogen and phosphorus [21]. Although the dilution value of these effluent discharges from aquaculture is considered high, its direct launch in the limnic environments can result in a chronic bioaccumulation and eutrophication, which can lead to an excessive increase of phytoplankton, causing dissolved oxygen deficit at night and possible death of local organisms [22]. Thus, the advancements in aquaculture simultaneously increase the concern of environmental agencies and societies with the environmental impact generated by it.

The main negative impact of effluents from aquaculture activities on aquatic ecosystems is the increase of nitrogen and phosphorus concentrations in the water column and accumulation of organic matter in the sediments [22]. In addition to effluent produced by natural processes, nutrient enrichment, feces, and unconsumed feed, chemical residues are also released, which are used in disinfection, pest and predator controls, disease treatments, hormones to induce reproduction, and sexual reversal beyond anesthetics for transport [23].

Not all breeding techniques have negative environmental consequences, since many of them are beneficial when environmental management is effective and socioeconomically sustainable [24]. As a positive impact of aquaculture, there are related consortiums between aquaculture and agriculture (irrigation and farming), rearing of tadpoles along with ornamental fish [25] or integrated systems of multiple uses such as recreation, gastronomy and rural tourism.

Another positive aspect is the maintenance of fish stocks in the sea and rivers, protecting and conserving endangered species, the use of industrial and domestic (treated) effluents in the enrichment of fish farms, or the coupling of a hydroponics system together with the residues of fish farming itself. It is also important to highlight the opportunities for new economic and working sources in the river basin [23] as a positive social and economic aspect.

The increasing growth of the animal industry has forced the reduction in the effects of intensive production systems on the environment. The concern with amino acids up to recent animal response considerations was restricted for maximizing production efficiency but little or no attention was given for reducing nitrogen excretion [26]. But now, there is a great concern about modern aquaculture's response to intensive production systems—the sustainability and environmental impact or pollution caused by them. In these systems, the density of fish per volume of water is very high, needing high rations and large amount of feed consisting of ingredients of high digestibility and palatability in order to produce minimum residue from feed wastage and stop high excretion of phosphorus and nitrogen [27].

The development of nutritious and environmentally and economically viable rations depends on the knowledge regarding the alimentary habits and nutritional requirements of the species reared [27]. Being an alternative production in the agriculture sector, frog culture can be placed among aquaculture activities that are gaining importance in the natural scenario [28], as the natural populations of frogs in Asia are decreasing due to environmental contamination and uncontrolled capture [29].

3. Importance of water quality for raniculture

The quality and cleanliness of water used in production of aquatic organisms are essential factors for the success of these programs. The frogs leave their excreta, and skin remains in water due to constant changes. It is imperative to constantly renew the water and clean the tanks and bays, and such care is necessary to prevent diseases and mortalities.

Amphibians such as bullfrog need water with specific physical and chemical quality. Parameters such as pH, electrical conductivity, total alkalinity, total hardness, ammonia, nitrite, nitrate, phosphorus, chlorides, iron, and especially oxygen must be measured before starting a breeding. These parameters are the most important indexes that characterize the quality of water. For breeders who are starting the activity, it is of the utmost importance to understand these variables.

All animal or vegetable life forms breathe in inhaling oxygen and exhaling carbon dioxide. When an aquatic environment is polluted with organic matter, the consumption of O_2 (respiration) exceeds beyond the acceptable levels and a decrease occurs in its available concentration. If the imbalance persists under anaerobic conditions (without oxygen), fish and most other animals will be unable to exist and will die. Oxygen allows aerobic (oxygen-using) bacteria to be more efficient decomposers than anaerobic (non-oxygen) bacteria, reducing decomposing organic matter in the water without leaving harmful odors.

When large quantities of organic material are discharged into rivers, for example, a population explosion of decomposing bacteria occurs. By "breathing," oxygen depletion occurs and the water becomes anaerobic or septic. The aerobic bacteria then facilitate the anaerobic bacteria, which produce hydrogen sulfide gas that has an extremely unpleasant smell and affect the aquatic life.

Aquatic ecosystems are dynamic, and even in tanks with small volumes of water, the physical and chemical parameters interrelate and are dependent on one another. The level of dissolved oxygen in water varies with temperature and atmospheric pressure. The dissolved oxygen content is about 9.08 mg L^{-1} at sea level and a temperature of 20°C, while this concentration rises up to 10.07 mg L^{-1} of oxygen, if the temperature drops to 15°C, indicating them the two closely interrelated factors.

The behavior of several other parameters occurs in the same way. Thus, it is not enough to know only one parameter or strictly follow the literature. The physical and chemical evaluation of water must be analyzed together, taking into account all factors.

3.1. Variables of interest in water quality control

pH (hydrogenation potential): It is the ratio between concentrations of hydrogen ions (H^+) and hydroxyl ions (OH^-), that is, acidity or alkalinity. It has a scale of 0–14, with pH 7 being neutral, where H^+ and OH concentrations are the same. Below 7, the pH indicates acidity, and above 7, it indicates alkalinity. The most responsible for its variation is the carbonic acid, which is originating from the carbon dioxide produced by phytoplankton during photosynthesis, where in excess, it renders the pH acidic and vice versa.

Electric conductivity: It is determined by the presence of dissolved substances that dissociate into anions and cations. It is the ability of water to transmit electric current. Practically for aquatic organisms, the higher the conductivity the more charged the system will be.

Total alkalinity: It indicates the concentration of carbonate and bicarbonate salts in water. It has a function of water buffering, that is, it maintains the pH stable, besides participating in the carapace formation of some species of plankton. Carbonates and other salts react with carbonic acid, neutralizing their action.

Total hardness: It indicates the concentration of metallic ions mainly the ions of calcium (Ca^{2+}) and magnesium (Mg^{2+}) present in water. It is expressed in $CaCO_3$ equivalents. The values of total hardness are practically associated with alkalinity. It also potentiates the toxicity of various chemicals.

Ammonia, nitrite, and nitrate: Ammonia produced due to the excretions of aquatic organisms and bacterial decomposition of the organic material in water is divided into toxic ammonia (NH_3) and ammonium ion (NH_4). Through bacterial oxidation (nitrosomonas), the ammonia is transformed into nitrite. Then, nitrite is oxidized by bacteria of the genus *Nitrobacter* to nitrate. The denitrifying bacteria transform nitrate into nitrogen by completing the cycle. Generally, ammonia and nitrite are the toxic forms (depending on pH and temperature). Nitrate is not toxic.

Phosphorus: It is a nutrient with low concentration in water but is one with the highest concentration factor in phytoplankton, followed by nitrogen and carbon. Its compounds constitute an important component of the living cell, especially nucleoproteins, essential for cellular reproduction. It is also associated with respiratory and photosynthetic metabolism. They occur mainly in the form of soluble phosphates and phosphate rock. Organic wastes, especially domestic sewage, contribute to the enrichment of water with this element.

Iron: Among the physical and chemical parameters of water, iron is the one that most frequently makes impossible the implantation of a commercial raniculture. This metal when in high concentration causes tadpoles mortality due to its chemical toxicity. It is sometimes possible to remove the iron from water through its oxidation (Fe_3—colloidal), in other words, introducing oxygen in the medium and aeration.

4. Water quality indices found in commercial farms for the cultivation of tadpoles

The excretion of the animals (feces and urine) results in ammonia-based compounds. Ammonia is extremely toxic when in large quantities and is converted into nitrite and nitrate by the action of nitrifying bacteria. Nitrite is also a toxic compound. It can oxidize hemoglobin in animals' blood; thus, converting it into methaemoglobin, a molecule incapable of carrying oxygen. This transformation process of ammonia (NH_3—toxic) in nitrite (NO_2—toxic) and then in nitrate (NO_3—toxic only in high quantities) is called denitrification and occurs depending on the temperature, pH, and oxygen of water. This reaction is one of the most common causes of mortality in tadpole tanks but it can also be easily avoided by taking basic precautions such as controlling the amount of food offered, constant and efficient oxygenation, water renewal, and regular cleaning.

The changes in water are related to values that allow classifying water by its degree of contamination, origin or nature of the main pollutants, and their effects to characterize cases of loads or peaks of concentration of toxic substances and to evaluate the biochemical balance necessary for maintenance of aquatic life. In other words, the farmer observing the water of his tanks daily can infer and/or perceive its state. However, not even the experience gained over the years will spare the farmer from regular observing/monitoring the water of his tanks.

In the literature, we find little information available on the ideal water quality for raniculture. Many concepts and values come from other types of aquaculture animals. Thus, a gap occurs when the farmers apply this information in commercial ranks, which is a more practical activity. Another way to aid in the elucidation of this process is to conduct aquatic research regarding the impact of water quality on frog culture on large-scale laboratory tests in order to provide more accurate and practical information to the frog culturists.

Following are some data collected in the field from observations made in commercial ranks in Brazil (**Table 1**), which are demonstrated.

Analyzing the above mentioned variables, it is verified that there is a very great similarity between fish and amphibians with respect to the physical and chemical parameters of the water. It has been noted that bullfrog tadpoles require less oxygen than fish. This is one of the reasons why many amphibians are known as "homebodies," that is, animals do not migrate from their place of origin when environmental conditions become adverse.

Normally, higher production of tadpoles is carried out in hot periods due to better adaptation and development of animals at an average temperature of 26°C [32]. The pH values did not show large changes among the frog farms, normally presenting average values of 7.0 for water supply and effluent, respectively. Thus, it remains within the standards recommended by the Brazilian resolution (pH between 6.0 and 9.0 for breeding waters of this species).

It is common to observe decrease in the dissolved oxygen content in the effluent than the water supply, but ideally, it is above the minimum required by the legislation (5.0 mg L^{-1}). As far as the ideal flow rate for spinning water is concerned, it is ideal that the total tank volume

Parameters	Desirable values	Values observed
pH	6.5–7.0	6.0–8.0
Oxygen	0.7–6.0 mg L^{-1}	2.0–6.0 mg L^{-1}
Ammonia (NH_3)	Up to 0.5 mg L^{-1}	Up to 0.7 mg L^{-1}
Nitrite (NO_2)	Up to 0.5 mg L^{-1}	Up to 1.0 mg L^{-1}
Nitrate (NO_3)	Up to 1.0 mg L^{-1}	—
Hardness	Up to 40 mg L^{-1}	10–80 mg L^{-1} $CaCO_3$ (most frequent)
Alkalinity	Up to 40 mg L^{-1}	10–80 mg L^{-1} $CaCO_3$ (most frequent)
Chloride (Cl_2)	Up to 7 mg L^{-1}	—
Chlorine (C_l)	0.02 mg L^{-1}	Up to 1 mg L^{-1}
Fluoride (F_2)	Less than 1 mg L^{-1}	—
Iron	Up to 0.3 mg L^{-1}	Up to 1 mg L^{-1}
Orthophosphate (PO_4)	Less than 0.3 mg L^{-1}	
Electrical conductivity	—	Less than 150 μS cm^{-1}

Adapted Ferreira et al. [30] and Hailey et al. [31].

Table 1. Physico-chemical characteristics of water observed in commercial bullfrog tadpole (*Lithobates catesbeianus*) farms in Brazil.

is renewed at least once a day in small tanks. Some authors have already described some flow rates found in frogs: Borges et al. [33] found 0.11 L s^{-1} and for the exit water, 0.08 L s^{-1}; and Pereira et al. [34] found the value of 0.064 L s^{-1}.

For electrical conductivity, there is no established standard for tadpoles. We find literature that demonstrates values of 30 to 200 μS cm^{-1} for incoming water and runoff water [35–37]. Mercante et al. [38] working with sequential nurseries of semi-intensive fish production observed a mean variation between 46 and 113 μS cm^{-1} in the nurseries (**Table 1**). The same author states that when the values are high, they indicate a high degree of decomposition and the inverse (reduced values) is intense primary production (phytoplankton).

The maximum limit for turbidity in effluents, according to Brazilian legislation, is 100 NTU (nephelometric turbidity units). According to Borges et al. [33], in a study with tadpoles, the maximum value obtained for incoming water was 20 NTU and for output was 26 NTU, thus remaining within the allowed range. Sipaúba-Tavares [39] argues that high levels of turbidity may be related to the presence of clays and dissolved or colloidal organic matter. In the same study by Borges et al. [33], ammonia and nitrate remained well below the limit established by legislation during the experimental period. However, from collection 2 (31 days of experiment), the concentrations were higher in the outlet water, with mean values of 0.40 and 0.49 mg L^{-1}, respectively, evidencing the organic matter decomposition and the rapid nitrification process due to the good oxygenation of the tanks.

5. Qualitative and microbiological characteristics of raniculture effluent

The main problem observed in raniculture as already observed in aquaculture is water enrichment (eutrophication) caused mainly by inadequate food management, which raises nutrient concentrations and modifies the environmental conditions of the farm [40]. The eutrophication process is a frequently observed problem and suggests the need for effluent studies and management techniques focused on ecological and specific aspects of these systems, favoring a lower impact of the effluents in the receiving water bodies [41].

In raniculture, commercial feed is usually used for carnivorous fish with average levels of 45% crude protein [42]. In addition to the increase in the amount of feed offered according to the growth of the animals, there is also a change in the use of this feed in each developmental stage of the animal [43]. The bullfrog produced in "anfigranja system" presents low protein efficiency of the commercial diet elaborated based on the requirements of carnivorous fish, which is used for feeding in most Brazilian ranicultures [44]. In addition, there is a significant loss of feed nutrients to the environment, resulting in the degradation of water quality [45].

A compound resulting from the catabolism of proteins is ammonia; therefore, the control of food quantity and quality, as well as adequate flow of water, is of fundamental importance for the maintenance of a good artificial breeding system [46]. In breeding systems for aquatic organisms, food introduced into the water and ammoniacal nitrogen fertilizers, such as ammonium sulfate, ammonium nitrate, phosphates, and urea, contribute to the increase of ammonia concentration in water [47]. The diet formulation should eliminate high levels of phosphorus and nondigestible components, testing the minimum necessity to grow a certain species [48]. Thus, the quality and quantity of food should be controlled for the sustainability of the harvesting system [49], since a relationship has been observed between commercial feed utilization and the water eutrophication process [50]. If the aquatic system is in imbalance, it becomes propitious to the development of diseases, compromising the sanitary sanity of that place.

Ectothermal animals such as bullfrog often serve as carriers for etiological agents by contact or as carriers when agents are ingested. When they are infected, it may result in an imbalance in the population of the pathogenic microorganisms in the environment, changes in the physical, chemical, and biological quality of water, immunological deficiency of these organisms, and unnecessary stresses [51].

The "Biological Institute of São Paulo, Brazil" has identified the presence of *Escherichia coli* bacteria in a very high load in diseased frogs (secondary and opportunistic infection) in a breeding system where the frogs fed larvae, and in the examination of feces of pigs used as substrate of larvae, an extremely high amount of this enterobacterium was reported [52].

E. coli is found in sewage, effluent, natural waters, and soils that have received recent fecal discharge, being inactivated. When ingested, it becomes active and pathogenic, but it is most used as indicator of fecal pollution in the environment and food, appearing in fresh and poor fish, frogs, mollusks, and shrimp [51].

Bacteria from the coliform group are also found in soil and vegetables, some with a certain ability to remain and even multiply in humid environments with high levels of organic and inorganic nutrients. Organisms in the coliform group may be introduced into water and food from nonfecal sources, such as plants and individual transporters, already polluted (lack of hygiene or sanity), such as contact with other animals or humans, even without release or direct contact with their excreta [51]. Gray et al. [53] suggest that frogs in metamorphosis or infected "froglets" may continue to release this pathogen, infecting other breeding members and eventually contaminating water sources. In addition, they can transport terrestrial pathogens to adjacent aquatic systems. It is possible that bullfrog adults may serve as suitable hosts for *E. coli* in stagnant aquatic systems. Another care that must be taken in raniculture is during the process of metamorphosis that is directly related to the reduced immunocompetence.

It has been reported that this microorganism can survive in fluvial waters for up to 27 days. Thus, persistence of *E. coli* in aquatic environments should be maintained through periodic contributions from primary reservoirs (homeothermic animals) or excreta contamination of wild host animals. The results of Borges et al. [33] suggest that bullfrogs can function as spill-over reservoirs of *E. coli* and thus contribute to its persistence in aquatic environments. In addition, since tadpoles in metamorphosis are capable of dispersion, they may play a role in the pathogen epidemiology [53]. **Table 2** shows the values of the biotic and abiotic variables of different types of aquatic cultures.

DO, dissolved oxygen; EC, electrical conductivity; BOD, biochemical oxygen demand; COD, chemical oxygen demand; MLN, most likely number.

Variables	**Pisciculture**	**Carciniculture**	**Tadpole**
	Macedo and Sipaúba-Tavares [54][1] **Sipaúba-Tavares et al. [55][2]** **Macedo et al. [56][3]**	**Henry-Silva and Camargo [57][1]** **Pistori et al. [58][2]** **Keppeler [59][3]**	**Borges et al. [45]**
EC (μS cm)	96[2]	70[2]	74
DO (mg L^{-1})	8.40[2]	4.63[2]	6.15
P-total (mg L^{-1})	0.25[2]	0.29[1]	1.88
Nitrate (mg L^{-1})	0.13[2]	0.62[3]	0.68
Ammonia (mg L^{-1})	0.10[2]	0.13[3]	0.82
BOD (mg L^{-1})	7[1]	7[3]	12
COD (mg L^{-1})	18[1]	—	51
Escherichia coli (MLN/100 ml)	4 × 10^6[3]	—	1.3×10^3

Table 2. Comparison between the values of biotic and abiotic variables found in effluents from different aquaculture activities.

6. Water quality indices in commercial farms for terrestrial frog cultivation

Frogs are water-dependent organisms, thus, for the elimination of excreta, controlling their body posture, respiration, reproduction, protection and safety makes the quality of water extremely important in breeding times [60]. The quality of water used in the production of aquatic organisms is one of the essential factors for the success of these enterprises. In raniculture, postmetamorphic animals leave their excreta and skin remains from constant changes in water. Therefore, it is important to constantly renew the water and clean the tanks and bays. Such care is essential for the prevention and prophylaxis of diseases, because when a disease sets in, mortality is certain [30].

The water for use in commercial frog farms should be of good quality, without fecal coliforms, heavy metals, and iron, with neutral pH, being preferably of spring or artesian well. It is recommended to select places with higher ambient temperatures for its rearing, since the frogs are ectothermic animals, presenting a more accelerated growth in higher temperatures. To maintain the quality of the breeding place, the water used in the farm must come from its own source and protected, do not receive polluting load of any kind, and have its reservoirs protected and cleaned regularly [61].

Raniculture projects should include knowledge of local hydrography and concern for the rational use of water, mainly to reduce impacts on water resources. Particular attention should be paid to the construction of projects in ecologically sensitive areas of importance to environmental preservation, such as permanent preservation areas. According to the Brazilian Institute of Fisheries, the flow necessary for the installation of commercial frog farms of 500 m^2 is 0.5 L s^{-1} ABRAPOA [62], resulting in an amount of 43,200 liters per day. However, this volume should not be discarded in the receiving water body without prior treatment.

The quality of the effluents must be periodically monitored, and the projects must provide for the installation of a treatment system for these effluents. Efforts should be made to increase the feed efficiency of the animals in order to ensure the reduction of the waste loads generated by the activity. The adoption of measures to reduce and eliminate the chances of diseases with preventive actions (sanitary management) and the maintenance of efficient and sustainable population densities is also important.

The main negative environmental impact observed is related to the degradation of the unconsumed feed, releasing to the water the nutrients and increasing the concentration of nitrogen and phosphorus. The excretion of animals (urea) is also released into the water and results in increased ammonia concentration. Remains of skin contribute to increase the amount of organic matter (total solids). Due to the nitrification processes that take place inside the bays, the amount of dissolved oxygen decreases drastically and reaches the anoxic levels.

The quality of the breeding place is based on the application of good animal breeding practices, in which the factors such as the technical knowledge about raniculture, ideal soil qualities (space), water in quality and quantity, constant, trained, and responsible workforce and projects that contemplate an economic planning should be properly addressed. The frog culture presents

Limnological variables	Bullfrog culture	Shrimp culture	Fish culture
	Borges et al. [45]	Henry-Silva and Camargo [57]	Sipaúba-Tavares et al. [55][1] Henry-Silva and Camargo [63][2]
Temperature (°C)	28.2	26.5	25.7[1]
pH	7.2	8.1	7.2[1]
Dissolved oxygen (mg L^{-1})	1.23	5.10	5.40[1]
Conductivity (μS cm^{-1})	249	68	62[1]
Turbidity (NTU)	66	62	26[2]
Total phosphorus (mg L^{-1})	6.09	0.23	0.20[1]
Ammonia (mg L^{-1})	6.94	0.02	0.07[1]
Nitrate (mg L^{-1})	2.37	0.16	0.16[1]

Table 3. Comparison of the limnological characteristics of effluent from frog, shrimp, and fish cultures.

higher levels of dissolved nutrients, mainly concentrations of phosphorus, ammonia, and conductivity, as compared to other aquaculture activities, such as fish and shrimp cultures (**Table 3**).

Borges et al. [45] (2012) concluded that the management adopted in ponds of frog growth positively changed the quality of water. In contrast to other cultures of aquatic organisms (fish and shrimp), effluent from frog culture has a greater potential to cause eutrophication in receiving bodies of water. Best aquaculture practices (BAPs) should also be recommended for frog culture in order to avoid water pollution and contamination of animals (food biosecurity).

Mercante et al. [38] evaluated the mean concentrations of total phosphorus (TP) and flow in the water (inlet and outlet) and the load produced per day for bullfrogs and compare their results with other aquatic production systems (**Table 4**).

	Inlet			Outlet			
	Concentration	Flow	Load	Concentration	Flow	Load	
	(mg L^{-1})	(L s^{-1})	(g day^{-1})	(mgL^{-1})	(L s^{-1})	(g day^{-1})	
Bullfrog farming	0.03	0.06	0.21	0.19	0.06	14.3	Mercante et al. [38]
Bullfrog farming	0.07	0.03	0.18	6.09	0.02	11.57	Borges et al. [45]
Tilapia farming	0.42	2.76	9.7	2.45	2.76	49.1	Pereira et al. [64]
Trout farming	72.26	40.61	233.33	99.69	40.61	343.67	Moraes et al. [65]

Table 4. Mean concentrations of total phosphorus (TP) and flow in water (inlet and outlet) and the load produced per day in different systems of animal production.

Mercante et al. [38] concluded that a constant renewal of water in the breeding bay is necessary to avoid the toxic effects in bullfrogs. However, it can promote higher nutrient loads. In order to improve the effluent quality and to reduce the nutrient load, in addition to effluent treatment, management options such as (a) flow maintenance and density reduction of animals and (b) maintain flow and density storage with better control of food supply, quality, and digestibility are proposed.

7. Usage of "wetland" in effluent treatment

Effluent treatment systems generated in aquaculture farms can be deployed using wetlands constructed to remove nutrients and improve water quality [66]. The aquatic macrophytes that are used in biological treatment systems, such as constructed wetlands, function as a bio-filter, improving the environmental conditions of production nurseries [67].

The constructed wetlands are tools used for the treatment of waste in aquaculture whose physical, chemical, and biological processes together with the local climatic conditions can improve the effluent quality [57, 68]. The importance of the implementation of these systems for the treatment of fish farming effluents in Brazil is due to the fact that many producers release water directly into natural streams and rivers [69], and the organic load exceeds its capacity of support and resilience [70].

The disposal of the effluent in the soil together with the presence of microorganisms, aquatic macrophytes, and solar energy results in the production of biomass and chemical energy, removing the polluting load. This is a system artificially designed to use aquatic plants (macrophytes) on substrates (such as sand, soil, or gravel), where the occurrence of biofilms with diverse populations of microorganisms treats wastewater through biological, chemical, and physical processes [71].

The conversion of ammoniacal nitrogen into wetlands is mainly due to two basic factors: the assimilatory process of microorganisms and macrophytes present in the systems and nitrification due to the transfer of oxygen from atmospheric air through the leaves of the macrophytes that through the aerenchyma permits the distribution of oxygen to the rhizomes and roots of plants [72].

The phosphorus present in the wastewater is generally phosphate and its removal in wetlands is controlled by the biotic and abiotic processes. The removal occurs due to the use of phosphorus by plants, periphyton and microorganisms, sedimentation, adsorption, precipitation, and exchange processes between the substrate and the water layer that remain in the system [73].

According to Travaini-Lima and Sipaúba-Tavares [68], the removal of phosphate compounds is associated to the hydraulic flow of the system, retention time, and the macrophyte species used, and species such as *Cyperus giganteus*, *Typha domingensis*, and *Eichhornia crassipes* are resistant and highly effective plants for subtropical regions. Sipaúba-Tavares and Boyd [74] verified that the wetland installed in an aquaculture farm containing only the aquatic plant *E. crassipes* presented efficiency in the removal of nitrogenous and phosphate compounds, improving the water quality of the effluent and also confirming that the system of biofiltration can be applied in shallow water channels.

The reduction of thermotolerant and total coliforms occurs due to the combination of physical, chemical, and biological factors. Physical factors include filtration through the plants, biofilm fixation on the substrate and on the macrophytes, and sedimentation. The chemical factors involve oxidation, biocidal effect resulting from the material excreted by some macrophytes, and adsorption of the organic matter. The biological mechanism includes production and effusion of chemical substances in the environment, which prevent the development of other organisms (antibiosis) and predation by nematodes and parasites, bacterial lysis, and inactivation [75].

Effluent treatment in aquaculture is one of the main factors in systems for breeding aquatic organisms that improve the water quality, avoiding or minimizing eutrophication in the receiving body. In addition, it is important to note that the use of nitrogen and phosphorus as a source of nitrogen and phosphorus in the aquatic ecosystem is a very important factor.

The importance of using biofiltration in aquaculture in different locations around the world is to ensure the development of this enterprise. The production of aquatic organisms varies from country to country, directly influenced by climatic and edaphic factors, as well as the population habit of each site. The incorporation of effluent treatment mechanisms of aquatic organism breeding systems depends on the economic conditions of the enterprise, the degree of pollution of the effluent, the cultivated species, and the ecological management employed [76].

Technology transfer should be stimulated toward the use of water or even in the treatment systems for the production of aquatic organisms, as a way of minimizing the impacts caused by aquaculture, being located and identifiable as possibilities of techniques for the mitigation of waste [77].

Integrated management of the system for the establishment of built wetlands and organisms results in significant productive and environmental gains. The cost of constructing the wetland system is similar to the cost of building stabilization ponds. The advantage of using wetland is the quality of the effluent, which can be used in crop irrigation, and the macrophytes can be used as material for green fertilization in agriculture, since they contain the nutrients withdrawn from the water stored in their biomass [78].

7.1. Use of "wetland" in raniculture

Raniculture or frog culture, as well as any other aquaculture practice, requires a large volume of water and produces effluent with high organic load and can be a significant source of local environmental impact. The main characteristics of this type of effluent are the high concentrations of dissolved nutrients, mainly ammonia and phosphorus, high electrical conductivity, and low dissolved oxygen concentration, as compared to effluents from other aquaculture practices [45].

Due to the need to treat the effluent from aquaculture practices, the use of "wetlands" constructed in a growing way in the country has been studied to minimize the impacts produced by these aquatic organisms [63, 66, 68, 69]. The aquatic plants extract or take out nutrients and other substances from the surrounding water which are necessary for their development, besides requiring low capital, low operating cost and versatility in the removal mechanism than conventional treatments. Sipaúba-Tavares and Braga [69] are also important in the removal of nitrogen, phosphorus, biochemical oxygen demand, thermotolerant solids, and coliforms [79].

Studies on the treatment of raniculture effluents are still scarce, and however, the aim of the wetlands is to avoid the degradation of the water quality of receiving reservoirs. Keeping this in view, Borges and Sipaúba-Tavares [80] constructed a wetland for the treatment of effluent from bullfrog breeding and evaluated the efficiency of removal of nutrients and thermotolerant coliforms in two phases of the fattening period (post metamorphosis). The total area and volume of the "wetland" were 14.2 m^2 and 2.2 m^3, respectively. It was 23 m long and composed of three boxes, the first with 0.51 m^3, the second 0.72 m^3, and the third 0.93 m^3, connected by channels with surface water flow, according to **Figure 1**. The first box was used only for sedimentation of solid residues, the second one filled with *Eichhornia crassipes*, and the third with *Typha domingensis* and *Cyperus giganteus* planted at the density of 5 m^2 plants (**Figure 1**).

In the study by Borges and Sipaúba-Tavares [80], water harvesting occurred in two distinct phases, phase I corresponded to the period of highest stock biomass (IF) between July and October 2012, when the animals weighed on average 276 g and the other in phase II (FII) between December 2012 and March 2013, where the animals weighed on average 29 g. The total biomass of animals in the bays in phase I was 351 kg and in phase II 218 kg.

Borges and Sipaúba-Tavares [80] found the following values for phase I: limnological variables such as pH and dissolved oxygen (DO) were higher ($p < 0.05$) at the exit and turbidity, and NO_3, NO_2, total suspend solids (TSS), and thermotolerant coliforms (TC) were higher ($p < 0.05$) at the entrance to the wetland (**Table 1**). In phase II, the values of the limnological variables such as turbidity, NO_2, NO_3, SST, and TC were higher ($p < 0.05$) at the entrance and DO and NH_3 were higher ($p < 0.05$) at the wetland exit (**Table 1**). In both phases, it was possible to observe an increase of DO in the water leaving the wetland in relation to the input water (**Table 5**).

In phase I, the highest values of NH_3, PT, SST, and TC occurred at the entrance of the wetland, and in phase II, the highest values of NO_2, total solids dissolved (TSD), biochemical oxygen demand (BOD), and chlorophyll occurred. NO_2 was approximately twice as high at the water inlet in the wetland of stage II, and NH_3 and STS were also higher at the entrance in phase I (**Table 5**).

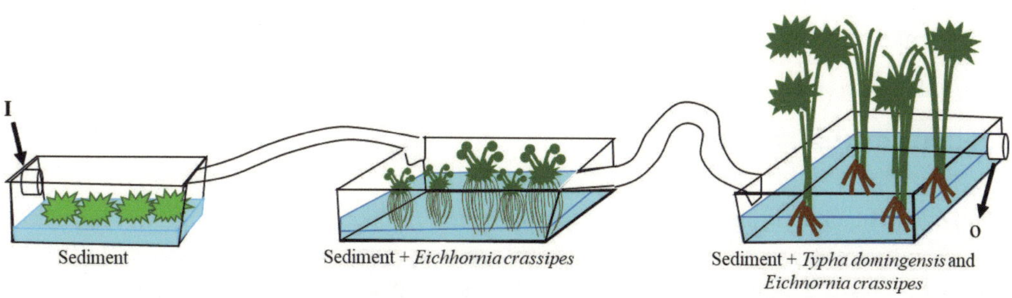

Figure 1. Layout of the wetland constructed in the frog breeding sector of the aquaculture center located at the State University of São Paulo (UNESP), Jaboticabal, SP, Brazil. I = entry point where the wastewater of the bullfrog production system entered the built marsh; O = the exit point where treated wastewater left the constructed wetland entering directly into a fish pond, adapted of Borges and Sipaúba-Tavares [80].

Variables	Phase I		Phase II	
	Entrance	Exit	Entrance	Exit
Temperature (°C)	28.0 ± 1.0	27.3 ± 1.0	27.9 ± 0.9	26.4 ± 0.9
pH	7.1 ± 0	7.3 ± 0.1	6.7 ± 0.3	6.9 ± 0.3
EC (μS cm^{-1})	182.9 ± 22.0	181.7 ± 19.9	162.6 ± 3.6	169.0 ± 5.4
Turbidity (NTU)	5.8 ± 1.6	2.9 ± 2.4	8.2 ± 3.4	2.3 ± 0.9
Alkalinity (mg L^{-1})	104.0 ± 2.3	108.4 ± 2.9	103.5 ± 1.6	104.9 ± 2.8
DO (mg L^{-1})	1.9 ± 0.3	2.6 ± 0.2	1.6 ± 0.4	2.9 ± 0.6
NO_3 (μg L^{-1})	82.7 ± 45.7	37.2 ± 14.4	81.8 ± 12.6	22.9 ± 18.5
NO_2 (μg L^{-1})	9.4 ± 3.6	5.3 ± 1.0	19.7 ± 3.1	11.6 ± 5.0
NH_3 (μg L^{-1})	2.785 ± 1.219.6	3.590 ± 955.9	1.388 ± 217.2	1.863 ± 164.9
PO_4 (μg L^{-1})	271.1 ± 132.6	309.9 ± 94.9	242.1 ± 52.2	265.4 ± 24.6
TP (μg L^{-1})	570.5 ± 285.5	480.4 ± 192.9	361.5 ± 102.2	283.4 ± 61.3
TSS (mg L^{-1})	21.0 ± 12.3	7.6 ± 6.8	9.0 ± 7.0	1.0 ± 0.8
STD (mg L^{-1})	135.7 ± 23.1	153.2 ± 31.7	194.5 ± 70.4	111.4 ± 9.5
BOD (mg L^{-1})	76.5 ± 13.6	71.6 ± 18.7	93.0 ± 3.8	85.6 ± 7.8
TC (NMP.100 m L^{-1})	50.750 ± 32.623	3.475 ± 2.284	25.250 ± 16.257	3.325 ± 1.362
Chlorophyll *a* (μg L^{-1})	4.5 ± 3.7	6.0 ± 4.8	7.2 ± 5.2	5.1 ± 3.4

EC, Electrical conductivity; DO, Dissolved oxygen; NO_3, Nitrate; NH_3, Nitrite; PO_4, Orthophosphate; TP, Total phosphorus; TSS, Total suspended solids; TSD, Total solids dissolved; BOD, Biochemical oxygen demand; TC, Thermotolerant coliforms.

Table 5. Mean, minimum, maximum, and standard deviation for the limnological variables of the water entering and exiting the wetland in phases I and II.

The authors concluded that there was a significant improvement in the quality of waste water of the bullfrog production system. Therefore, the treatment used was adequate and could be used in commercial farms, with only a few adjustments being made to improve the efficiency of nutrient removal.

8. Final considerations

Raniculture like other aquaculture practices is facing problems that affect the water regime, impacting natural water sources and directly water quality, which are responsible for the successful breeding and production systems. Although there are few studies on water quality in raniculture, aquaculture has already a number of studies but still there is a lack of understanding of how these ecosystems actually function and interact with biotic and abiotic factors. Any sudden disturbance rapidly changes the water quality and consequently causing certain damages to the commercially important species. Thus, future studies should focus on the point to explain how different aquaculture systems function and how they interact with biotic and abiotic factors. These practices may provide excellent information to establish successful aquaculture systems with favorable environmental influences.

Acknowledgements

We thank the Brazil University (Univ. Brasil) for financial support.

Author details

Cleber Fernando M. Mansano[1]*, Luiz Sérgio Vanzela[1], Juliana Heloisa P. Américo-Pinheiro[1], Beatrice I. Macente[1], Kifayat U. Khan[2], João Batista K. Fernandes[2], Danila Fernanda R. Frias[1] and Marta V. De Stéfani[2]

*Address all correspondence to: clebermansano@yahoo.com.br

1 Brazil University, Fernandópolis, SP, Brazil

2 Aquaculture Center of UNESP—Universidade Estadual Paulista, Jaboticabal, SP, Brazil

References

[1] Oliveira EG. Ranicultura: Novos desafios e perspectivas do mercado. Ciência Animal. 2015;**25**(1):173-186

[2] Carraro KC. Ranicultura: Um bom negócio que contribui para a saúde. Revista FAE. 2008;**1**:111-118

[3] Food and Agriculture Organization of the United Nation–FAO. The State of World Fisheries and Aquaculture; 2008. Rome: FAO - Fisheries and Aquaculture Department; 2009. 146 p

[4] Figueiredo RBA. ranicultura no Brasil é renda certa para o produtor. Revista Eletrônica Nordeste Rural. 2005. 12 p

[5] Nascimento R, Mello SCRP, Seixas Filho JT. Manual prático para criação de rãs com reuso de água: girinagem e metamorfose. Rio de Janeiro: SUAM; 2013. 82 p

[6] Seixas Filho JT, Pereira MM, Mello SCRP. Manual de Ranicultura Para o Produtor. Rio de Janeiro: FIPERJ; 2017. 155 p

[7] Seixas Filho JT, Mello SCRP, Veiga RCA, Miranda RGB, Santos CAN. Efeito da granulometria da ração sobre o desempenho de girinos de *Rana catesbeiana*. Revista Brasileira de Zootecnia. 1998;**27**:224-230

[8] Tiago G, Gianesella SMF. Uso da água pela aquicultura: Estratégias e ferramentas de implementação de gestão. Boletim do Instituto de Pesca. 2003;**29**(1):1-7

[9] Phillips MJ, Beveridge MCM, Clark RM. Impact of aquaculture on water resources. In: Brune DE, Tomasso JR, editors. Aquaculture and Water Quality. Baton Rouge: The World Aquaculture Society; 1991. pp. 568-591

[10] Telles DA. Água na Agricultura e Pecuária. In: Rebouças AC, Braga B, Tundisi JG, editors. Águas doces no Brasil: Capital ecológico, uso e conservação. São Paulo: Escrituras Ed. e Distr. de Livros Ltda; 2002. pp. 305-337

[11] Watanabe T, Sakamoto H, Abiru M, Yamashita J. Development of a new type of dry pellet for yellowtail. Nippon Suisan Gakkaishi. 1991;**57**(5):891-897

[12] Cho CY, Hynes JD, Wood KR, Yoshida HK. Development of highnutrient-dense, low-pollution diets and prediction of aquaculture wastes using biological approaches. Aquaculture. 1994;**124**:293-305

[13] Valenti WC. Aquaculture for sustainable development. In: Valenti WC, Poli CR, Pereira JA, Borghetti JR, editors. Aquicultura no Brasil: Bases para um desenvolvimento sustentável. Brasília: CNPq/MCT; 2000. 399 p

[14] Valle CE. Qualidade ambiental: ISO 14000. 6th ed. São Paulo: Editora Senac/SP; 2006. 96 p

[15] Piedratha RH. Managing environmental impacts in aquaculture. Bulletin of National Research Institute of Aquaculture Supplement. 1994;**1**:13-20

[16] Gupta MV, Dey MM. A framework for assessing the impact of small-scale rural aquaculture projects on poverty alleviation and food security. FAO Aquaculture Newsletter. 1999;**23**:18-22

[17] Ayoleke OE. Effects of aquaculture pratices on water quality of pond-waste receiving water. Aquaculture. 2002;**33**(2):60-62

[18] Sipaúba-Tavares LH, Gomes JPFS, Braga FMS. Effect of liming managenment on the water quality in *Colossoma macropomum* ("tambaqui") ponds. Acta Limnologica Brasiliensia. 2003;**15**:95-103

[19] Boyd CE, Hulcher R. Best managements practices established for channel catfish farming in Alabama. AAES Highlights. 2002;**1**:1-4

[20] Silvert W. Assessing environmental impact of finfish aquaculture in marine waters. Aquaculture. 1992;**107**:67-79

[21] Boyd CE. Guidelines for aquaculture effluent management at the farm-level. Aquaculture. 2003;**226**(1-4):101-102

[22] Reddling T, Midlen A. The treatment of aquaculture wastewater: A botanical. Journal of Environmental Management. 1997;**50**:283-299

[23] Eler MN, Millani TJ. Métodos de estudos de sustentabilidade aplicados a aquicultura. Revista Brasileira de Zootecnia. 2007;**36**:33-44

[24] Stephens WW, Farris JL. A biomonitoring approach to aquaculture effluent characterization in channel catfish fingerling production. Aquaculture. 2004;**241**:319-330

[25] Nagata MK. Criação de girinos consorciada a peixes ornamentais. In: Simpósio Brasileiro De Ranicultura, 1, Ciclo De Palestras Sobre Ranicultura Do Instituto De Pesca, 2, São Paulo. Vol. 34. São Paulo: Instituto de Pesca; 2003. pp. 52-58

[26] Braga JP, Baião NP. O conceito de proteína ideal na formulação de ração para frangos de corte. Caderno Técnico de Veterinária e Zootecnia. 2001;**34**:29-37

[27] Portz L, Cyrino JEP, Martino RC. Effect of dietary protein and energy levels on growth and body composition of juvenile largemouth bass *Micropterus salmoides*. Aquaculture Nutrition. 2001;**7**:1-8

[28] Feix RD, Abdallah PR, Figueiredo MRC. Resultado econômico da criação e rã em regiões de clima temperado. Brasil. Informações Econômicas. 2006;**36**(3):70-80

[29] Teixeira RD, Pereira Mello SCR, Santos CAML. The World Market for Frog Legs. FAO/Globefish Research Programme. Vol. 68. Rome: FAO; 2001. 44 p

[30] Ferreira CM. A importância da água e sua utilização em ranários comerciais. Revista Panorama da Aquicultura. 2003;**13**(79):15-17

[31] Hailey A, Sookoo N, Mohammed A, Khan A. Factors affecting tadpole growth: Development of a rearing system for the Neotropical leptodactylid *Physalaemus pustulosus* for ecotoxicological studies. Applied Herpetology. 2006;**3**(2):111-128

[32] Ferreira CM, Pimenta AGC, Paiva Neto JS. Introdução à ranicultura. Vol. 33. São Paulo: Boletim Técnico do Instituto de Pesca; 2002. 15 p

[33] Borges FF, Stéfani MV, Amaral LA. Quality of the Effluents of Bullfrog Tadpole Ponds. Boletim Do Instituto de Pesca. Vol. 402014. pp. 409-417

[34] Pereira JS, Mescadante CTJ, Lombardi JV, Caruso NPP, Osti JAS, Miashiro Let al. Caracterização da qualidade da água, através de fatores abióticos, da entrada e saída de um sistema de produção de rãs (*Rana catesbeiana* Shaw). In: Congresso de Ecologia do Brasil; 8, 2007. Caxambu: Instituto de Pesca; 2007

[35] Mansano CFM, Stéfani MV, Pereira MM, Macente BI. Non-linear growth models for bullfrog tadpoles. Ciência e Agrotecnologia. 2012;**36**(4):454-462

[36] Mansano CFM, De Stéfani MV, Pereira MM, Macente BI. Deposição de nutrientes na carcaça de girinos de rã-touro. Pesquisa Agropecuária Brasileira. 2013;**48**(8):885-891

[37] Mansano CFM, De Stéfani MV, Pereira MM, Nascimento TSR, Macente BI. Morphometric growth characteristics and body composition of bullfrog tadpoles in captivity. Semina: Ciências Agrárias. 2014;**35**(5):2817-2830

[38] Mercante CTJ, Vaz-dos-Santos AM, Moraes MDAB, Pereira JS, Lombardi JV. Bullfrog (*Lithobates catesbeianus*) farming system: Water quality and environmental changes. Acta Limnologica Brasiliensia. 2014;**26**:9-17

[39] Sipaúba-Tavares LH. Limnologia aplicada à aquicultura. São Paulo: Funep; 1994. 72 p

[40] Piedrahita RH. Reducing the potential environmental impact of tank aquaculture effluents through intensification and recirculation. Aquaculture. 2003;**226**(1):35-44

[41] Macedo CF, Sipaúba-Tavares LH. Eutrofização e qualidade da água na piscicultura: Conseqüências e recomendações. Boletim do Instituto de Pesca. 2010;**36**(2):149-163

[42] Mansano CFM, Macente BI, Nascimento TMT, Pinto DFH, Pereira MM, De Stéfani MV. Digestibility of nutrients and energy in ingredients for bullfrogs during different phases of development. Aquaculture Nutrition. 2017;**23**:1368-1378

[43] Mansano CFM, Macente BI, Nascimento TMT, Pereira MM, Takahashi LS, De Stéfani MV. Amino acid digestibility of animal protein ingredients for bullfrogs in different phases of post-metamorphic development. 2017;**48**(9):4822-4835

[44] Pereira MM, Mansano CFM, Silva EP, de Stéfani MV. Growth in weight and of some tissues in the bullfrog: Fitting nonlinear models during the fattening phase. Ciência Agrotecnologia. 2014;**38**(6):598-606

[45] Borges FF, Amaral LA, Stéfani MV. Characterization of effluents from bullfrog (Lithobates catesbeianus, Shaw, 1802) grow-out ponds. Acta Limnologica Brasiliensia. 2012;**24**(2):160-166

[46] Pereira LPF, Mercante CTJ. A amônia nos sistemas de criação de peixes e seus efeitos sobre a qualidade da água: Uma revisão. Boletim do Instituto de Pesca. 2005;**31**(1):81-88

[47] Kubitza F. Qualidade da água na produção de peixes. Panorama da Aquicultura. 1998; **46**(8):35-41

[48] Sipaúba-Tavares LH, Santeiro RM. Fish farm and water quality management. Acta Scientiarum. Biological Sciences. 2013;**35**(1):21-27

[49] Boyd CE, Tucker C, Mcnevin A, Bostick K, Clay J. Indicators of resource use efficiency and environmental performance in fish and crustacean aquaculture. Reviews in Fisheries Science. 2007;**15**(4):327-360

[50] Simões FS, Moreira AB, Bisinoti MC, Gimenez SMN, Yabe MJS. Water quality index as a simple indicator of aquaculture effects on aquatic bodies. Ecological Indicators. 2008;**08**: 476-484

[51] Pádua HB. Informações sobre os coliformes totais/fecais e alguns outros organismos indicadores em sistemas aquáticos – aquicultura. 23ª Procuradoria de Justiça Criminal de Goiás, Caderno de Doutrina Ambiental; 2003. 20 p

[52] Braz Filho MSP. Criação de rãs. Curso/apostila; 1999. 29 p. mbraz@pescar.com.br, www.abrappesq.com.br, www.pescar.com.br/mbraz

[53] Gray MJ, Rajeev S, Miller DL, Schmutzer AC, Burton EC, Rogers ED, et al. Preliminary evidence that American bullfrogs (*Rana catesbeiana*) are suitable hosts for *Escherichia coli* O157:H7. Applied and Environmental Microbiology. 2007;**73**(12):4066-4068

[54] Macedo CF, Sipaúba-Tavares L. Variação de nutrientes e estado trófico em viveiros sequenciais de criação de peixes. Acta Scientiarum Animal Science. 2005;**27**(3):405-411

[55] Sipaúba-Tavares LH, Bachion MA, Colus DSO. Estudos limnológicos em três viveiros de criação de peixes com fluxo contínuo de água. Boletim Técnico do CEPTA, Pirassununga. 2006;**19**:35-47

[56] Macedo CF, Sipaúba-Tavares L, Amaral LA. Aspectos sanitários em viveiros de criação de peixes com disposição sequencial. Magistra, Cruz das Almas/BA. 2009;**21**(2):73-80

[57] Henry-Silva GG, Camargo AFM. Tratamento de efluentes de carcinicultura por macrófitas aquáticas flutuantes. Revista Brasileira de Zootecnia. 2008;**37**(2):181-188

[58] Pistori RET, Henry-Silva GG, Biudes JFV, Camargo AFM. Influence of aquaculture effluents on the growth of Salvinia molesta Influência do efluente de aqüicultura sobre o crescimento de Salvinia molesta. Acta Limnologica Brasiliensia. 2010;**22**(2):179-186

[59] Keppeler EC. Correlações limnológicas em viveiros de cultivo do camarão-da-amazônia *Macrobrachium amazonicum*. Biotemas. 2008;**21**(4):65-72

[60] Hipolito M. Manejo sanitário no cultivo de rã. p. 330-351. In Ranzani-Paiva MJT, Takemoto RM, Lizama MAP. Sanidade de organismos aquáticos. Varela, São Paulo. 2004. 426 p

[61] Hipolito M. Qualidade dos produtos da aquicultura – ranicultura. In: Silva Souza AT, editor. Sanidade dos organismos aquáticos no Brasil. Maringá, PR: Associação Brasileira de Patologia de organismos Aquáticos (Abrapoa); 2006. 387 p

[62] ABRAPOA – Associação Brasileira De Patologistas De Organismos Aquáticos. Ranicultura. Boletim Eletrônico. 2006 http://www.abrapoa.org.br/boletimeletronico/n13/index.html#oque

[63] Henry-Silva GG, Camargo AFM. Efficiency of aquatic macrophytes to treat Nile tilapia pond effluents. Scientia Agricola. 2006;**63**:433-438

[64] Pereira S, Mercante CTJ, Lombardi JV, Osti JAS, Miashiro L. Environmental characterization of pisciculture fishpond, São Paulo, Brazil. In: Proceedings Symposium Brasil-Japão – SBPN. Brazil Japan Researchers Association; 2008. pp. 1-5

[65] Moraes MAB, Caramel BP, Carmo CF, Tabata YA, Osti JAS, Ishikawa CM, et al. Dinâmica do fósforo em sistema intensivo de criação de trutas. In: Anais do Reunião Científica do Instituto de Pesca. São Paulo: Instituto de Pesca; 2013. pp. 263-265

[66] Millan RN, Sipaúba-Tavares LH, Travaini-Lima F. Influence of constructed wetland and soil filter systems in the dynamics of phytoplankton functional groups of two subtropical fish farm wastewaters. Journal of Water Resource and Protection. 2014;**6**:8-15

[67] Dias SG, Sipauba-Tavares LH. Physical, chemical and microbiological aspects during the dry and rainy seasons in a pond covered by macrophyte used in aquaculture water supply. Acta Limnologica Brasiliensia. 2012;**24**(3):276-284

[68] Travaini-Lima F, Sipaúba-Tavares LH. Efficiency of a constructed wetland for wastewaters treatment. Acta Limnologica Brasiliensia. 2012;**24**(3):255-265

[69] Sipaúba-Tavares LH, Braga FMS. Constructed wetland in wastewater treatment. Acta Scientiarum. Biological Science. 2008;**30**:261-265

[70] Santiago AF, Calijuri ML, Luís PG. Potencial para a Utilização de sistemas de wetlands no tratamento de águas residuárias: uma contribuição a sustentabilidade dos recursos hídricos no Brasil. Natureza e Desenvolvimento. 2005;**1**(1):29-39

[71] Sousa JT, Van Haandel AC, Da Silva Cosentino PR, Guimarães AVA. Pós-tratamento de efluente de reator UASB utilizando sistemas "wetlands" construídos. Revista Brasileira de Engenharia Agrícola e Ambiental. 2000;**4**(1):87-91

[72] Cooper PF, Green MD, Shutes RBE. Reed Beds and Constructed Wetlands for Wastewater Treatment. Buckinghamshire: WRC Publications; 1996. 206 p

[73] Reddy KR, D'Angelo EM. Biogeochemical indicators to evaluate polluant removal efficiency in constructed wetlands. Water Science and Technology, London. 1997;**35**(5):1-10

[74] Sipaúba-Tavares LH, Boyd CE. Macrophyte biofilter for treating effluent from aquaculture. In: Aquaculture Annual Technical Report. United States: Oregon; 2005

[75] Rivera F, Warren A, Ramirez E, Decamp O, Bonilla P, Gallegos E, et al. Removal of pathogens from wastewaters by the root zone method (RZM). Water Science and Technology. 1995;**32**(3):211-218

[76] Sipaúba-Tavares LHS. Uso racional da água em aquicultura. Jaboticabal: Maria de Lourdes Brandel – ME; 2013. 190 p

[77] Sipaúba-Tavares LH. Utilização de biofiltros em sistemas de cultivo de peixes. Informe Agropecuário. 2000;**21**(203):38-43

[78] Sousa JT, Van Haandel AC, Lima EPDC, Henrique IN. Utilização de wetland construído no pós-tratamento de esgotos domésticos pré-tratados em reator UASB. Engenharia Sanitária e Ambiental. 2004;**9**(4):285-290

[79] Maltais-Landry G, Maranger R, Brisson J, Chazarenc F. Nitrogen transformations and retention in planted and artificially aerated constructed wetlands. Water Research. 2009; **43**:535-545

[80] Borges FF, Sipaúba-Tavares LH. Treatment of bullfrog farming wastewater in a constructed wetland. Journal of Water Resource and Protection. 2017;**9**:578-589